Getting to Know a
— in School an

I am grateful for the patient help and wise advice given to me by Paul Black during the initial stages of this work, the inspiration of my many pupils and teaching colleagues, and the continual encouragement of John Ziman.

Getting to Know about Energy – in School and Society

Joan Solomon

RoutledgeFalmer
Taylor & Francis Group

LONDON AND NEW YORK

© J. Solomon, 1992

All rights reserved. No part of this publication may be reproduced, stored in a retrieval system, or transmitted, in any form or by means, electronic, mechanical, photocopying, recording, or otherwise, without permission in writing from the Publisher.

First published 1992
By RoutledgeFalmer,
2 Park Square, Milton Park, Abingdon, Oxon, OX14 4RN

Transferred to Digital Printing 2004

**British Library Cataloguing in Publication Data
is available on request**

**Library of Congress Cataloging-in-Publication Data
is available on request**

ISBN 0 75070 018 1 Case
ISBN 0 75070 019 X Paper back

Set in 10/12pt Times
by Graphicraft Typesetters Ltd, Hong Kong

Contents

	Preface	vi
Chapter 1	General Knowledge about Energy	1
Chapter 2	Schools of Thought on Children's Notions	18
Chapter 3	Children's Ideas on Energy	40
Chapter 4	Evidence from Talking	59
Chapter 5	Learning as an Extension of General Knowing	78
Chapter 6	The Formal Domain of Scientific Knowledge	98
Chapter 7	Teaching about Conservation	116
Chapter 8	The Social Uses of Energy	141
Chapter 9	The Defining Technologies of Energy	167
	References	189
	Index	195

Preface

This book, as its title suggests, is about education in and beyond the school. But it stretches the whole notion of education to a point where including the word in its title might have been misleading.

Several of its chapters do describe the school teaching and learning situation in considerable depth, quoting liberally from the questions and answers of my own and other pupils. Chapter 2 is almost completely devoted to describing different perspectives in science education research. Parts of chapters 4 to 7 give illustrated classroom accounts of learning about the transfer, conservation and degradation of energy which are used to explore and explain children's thinking on these topics. All these chapters also develop quite new ideas about related but broader educational issues ranging from the general aims of education in the public domain, to the operation of that most intimate form of teaching inquiry — action research in the classroom. There are passages which describe those tacit understandings which make a tenuous link between doing practical activities in the laboratory and learning science concepts. There are poems and drawings by younger children and arguments about moral dilemmas related to energy use from older school students.

Some chapters in the book, and considerable sections in many others, escape completely from the classroom into considerations of energy in language, literature, beliefs, technology, history and general culture. This is partly because their absence would make the school experience of our children quite incomprehensible, and indeed the main educational thesis developed in chapter 6 is devoted to the 'Two Domains' of knowledge — life-world, and abstract academic — that children carry in their heads. But wider considerations are even more important for their vital role in informal processes for getting to know about energy throughout life. School is only an episode in this longer view of learning, and schooling itself is laced through with meanings and experiences which could not possibly figure in any energy syllabus. Television, and current concerns about the *Public Understanding of Science* related to energy, are considered in chapters 4 and

Preface

8, and few could argue that school children are not involved in either of these topics. More ambitiously a theme is developed about general knowledge and common sense knowing in chapter 1 which is used in many contexts throughout the book.

The final chapter may seem, at first sight, to abandon science education as well as children's schooling. It develops a theme about energy technology in history and how it has affected people's ways of living and their values, in earlier times. Some of this has descended to us through common figures of speech. But the purpose of this line of thought is not retrospective: it is to make a little sense of the theme which is likely to loom largest in the next century's thinking about energy.

The wide scope of this theme presents almost insuperable problems of scholarship. The width of knowledge that would be required to make proper reference to the various bodies of literature involved — in cognition, social psychology, developmental psychology, education, philosophies of all sorts and varieties, sociology, the sociology of knowledge, technology, history, etc. . . . is quite out of my reach. I offer apologies for this — but only up to a point. More thoroughness in quoting references and making acknowledgements would certainly serve to protect the author against imputations of ignorance, but it would also place the work in a category of review literature which was not intended. The modern habit of writing multi-author books on a loose general theme addresses this problem in another way. While each contributor then supplies expertise and enthusiasm in a separate area, the result cannot easily be a complete thesis or narrative. The choice is between continuity of purpose and narrative with incomplete scholarship — or a more learned approach with the possibility of disjointed argument and reading. Probably there is no uniquely satisfactory solution to the problem. This book may speak to many audiences so while any one reader may find it regrettably deficient in her/his own speciality, they may find it more novel in another. It has taken on board only one topic, but carried it to many shores.

The topic — energy — is just right for such a wide-ranging investigation. It is thought to be the most difficult of all concepts to teach, and yet the term has re-invaded the everyday vocabulary from which it was borrowed by physicists less than two centuries ago, and stirred up a multitude of related ideas and folk beliefs. It is also high on the list of physicists' most fundamental ideas. Scratch any physicist and you will find an enthusiast for the energy concept. As a newly-graduated physicist just going into teaching, many years ago, I was certainly one of those. The charm has not worn off.

This book is not about higher education in the physics of energy, and yet, as the kitchen garden of all teachers, the university and polytechnic are fundamental to school learning about energy and so to the genesis of this book. The time-honoured method of teaching in colleges by means of lectures which are often monotonous and sometimes inaudible, may seem to have little of value to offer a prospective teacher. But, paradoxically, the private struggle to understand can be of enormous value. The personal

Preface

construction of knowledge, which will be a theme in several chapters, but particularly in chapter 2, is much harder when the presentation of knowledge is poor; but for those who remember the head-scratching difficulty of trying to understand lecture notes taken in the desperation of confusion, it makes a quite excellent background for thinking about teaching.

The first requirement is that the student's motivation to learn about energy is maintained. For most physicists interest in energy may have begun in school, but at university the spell often works wonderfully. The story of how the spectrum of energies in the beta-emission from a radioactive source suggested to Pauli that some other particle must also be formed *in order that* the all-important principle of Conservation of Energy would not be flouted, gives a good introduction to physicists' psychological commitment to the concept. For them contravention of the principle was so unthinkable that a strange new particle had to be thought into existence without electric charge or rest mass, but yet carrying spin. The steadfastness with which the physics community put their trust in a particle so tiny and elusive that it would, by their own calculations, travel through ten thousand million million miles of solid rock with less than a 50 per cent chance of encountering a single atom in its path, was impressive commitment.

Statistical mechanics also comes early in the undergraduate course and gives another valuable lesson for energy teachers. It uses the mathematical tricks learnt for dealing with the collisions of sizeable objects, on a multitude of point masses, and suddenly energy begins to assume a new dimension. When the energy laws are applied to crowds they take on the punter's feel for the probability of happenings. Energy changes, in this domain, were less about 'doing work' and more about statistical likelihood. These lectures took us students on one of our first journeys into the transmutation of meanings. Through statistical probability our understanding of the meaning of energy began to make contact with ideas of temperature and even time.

Forming links between concepts, like the use of simile to link descriptions, involves a rethinking at both ends of the chain of comparison. This is a topic which is also important in science education where analogies from the everyday world are used and stretched to fit physics theories. Few of us make any claim to an understanding of time, and that very ignorance allows the concept some freedom of imaginary movement as it sweeps up fragments of new meaning from poetry or from physics. But everyone thinks they know about temperature. A little girl must learn to dread the fire; the autonomous reflexes of the body ensure that sensorimotor learning about temperature precedes even the thinking about it. But there are many kinds of learning and each begins to grow in almost virgin territory. To understand the thermometer we leave behind the pain of fire. To understand the temperature of an assembly of rushing molecules in a gas we abandon the thermometer. Every time a student of physics learns in a new field they need to accept that, like a child opening a Russian doll, they are looking into a new reality where concepts with familiar names may not exactly match the

Preface

older ones they already have in mind. That also makes a good lesson for the intending teacher.

Other invariant concepts, like the mass of 'solid' matter and even the vectorial thrust of momentum, are too straightforward to reveal the direction of time flow. Such concepts are precisely and prosaically conserved — exactly the same in the past as in the future. But energy is both conserved and yet time-dependent in its organization. Energy is *not* the same after an event as it was before it, except in the simplest arithmetic sense. It is that difference, from order to disorder, from hot and cold to warm, from potentiality to action, or from action to spent quiescence, which spells out the passage of time.

University lectures on energy, in those days, made no pretensions to connect with contemporary values and fears, but science can neither be constructed nor learnt in a neutral spirit and a personal vacuum. We were taught, for example, about Lord Kelvin's attempts to calculate the age of the earth using no more than the nature of energy flow, and about his prediction of the slow 'heat death' of the universe. But we were the children of the Second World War and feared the recently dropped atomic bomb far more than Kelvin's ultimate running-down. The 1950s were a time of nuclear uncertainty punctuated by test explosions, in the earth's atmosphere, of ever more powerful bombs. New meanings have to resonate with societal and affective factors as well as with intellectual ones. For us, Kelvin's programme of destruction was too intellectually conceived and too gradual. Our lecturers never touched on social issues — this was long before the movement for including 'social relevance' in science education — they merely pointed to faults in Kelvin's mathematical scale of destruction. We smiled at the thwarted ambition of Victorian theoretical physics, and passed on to learn about the world of the atom itself, which seemed so very much more threatening.

We learnt that energy has its own discrete, almost atomic, structure. At the turn of the century contradictions in the physics of atomic interactions had driven Max Planck, rather reluctantly, to go beyond the comfortable confines of the classical notion of the continuity of energy. The very first indivisible entity, the first quantum, he encountered was that of 'action'. This was a small but strangely enduring quantity — the product of energy and time. The act of turning that over in our minds constructed yet another meaning, and also an apparent contradiction. When a transfer of energy takes place there is a calculable uncertainty in its time-scale. The smaller the amount of energy the greater is the vagueness in how long the exchange takes. That meant that the elementary meaning for energy, explained so clearly and didactically at school in terms of the strict measurement of force and displacement, was being superseded by another. Was it possible to believe in both meanings at the same time? For potential science teachers that demonstration of cognitive conflict was the best possible preparation.

Preface

This difficult movement from one domain of meaning to another matches the problems facing school pupils as they move from the everyday world of energy to that of school text-book physics. The daily struggle to understand why children find learning science so hard, teaches the teachers. From this they come to understand, if they are prepared to listen, more about the nature of learning and knowledge than can be given through half a dozen lectures about cognition and epistemology. The restless collection of children who tumble noisily into the classroom are the stuff of an invaluable learning laboratory. Years later, when I taught experienced teachers in a Master's course about the history and philosophy of science in terms of the long struggle to redescribe natural phenomena in a coherent way, I was often surprised by the range of useful understandings and insights which the classroom activity had given them.

Britain is fortunate in having an all-graduate force of dedicated school science teachers, despite the disincentives of salary and status. They still trickle into teaching, as I had done, with their heads full to the brim with science concepts. I watch with pleasure and some sympathy as the new PGCE entrants, still warm from the spell of undergraduate physics, set out to fire the young with tales of energy, and other mysteries. They are keen to stand in front of children and talk about science in a way that will make young eyes sparkle. This is the common 'Ancient Mariner' delusion, and it deceives the novice teacher badly. Few of us, if any at all, have the necessary 'glittering eye' to hold an audience: new teachers fail most often when they try to talk enthusiasm into their pupils from front-of-class. It has been my professional duty to deter student teachers from such didactic exposition with harsh advice about classroom discipline and laboratory management. (I always hope that I only half succeed!)

Some students leave teaching after the first scorching probationary year. Others stay on for a number of reasons — for pleasure in the company and comments of youngsters, because the challenge of teaching management catches their ambition, for the rare moments when they see the glint of real understanding and enjoyment in a new young physicist, and no doubt for many other reasons. Some of us may also learn that there is an intellectual challenge in teaching which is every bit as hard as a problem in advanced thermodynamics — understanding the pupils' own difficulties.

How do children learn? What does it mean for any of us to understand energy? No book could ever answer such questions. Instead, this one describes different kinds of understandings and how each one weaves its meanings into the central business of living and learning. It has to touch upon language and personality, on the ancient and the topical, on cognitive growth and personal values, on practical technology and abstract theory, on the world of citizens and the world of schools. Learning about energy is not something which only happens at school between the nine o'clock and ten o'clock bells twice a week during the autumn term. Energy is an idea with a multitude of meanings most of which thrive prodigiously out of school but are not halted at the playground gate. Language brings energy meanings into everything we do and think. This book is about energy and, as children say, 'energy is the source of all life'.

Chapter 1

General Knowledge about Energy

The Nature of General Knowledge

The place to start any search about understanding is in the general knowledge of the public. This is important for the study of learning not only because it is the foundation on which education has to build but because it will continue to flow strongly in daily conversations throughout life. Scientist or layman, expert or novice, we are united by our general knowledge. But before it is possible to think about the contents of general knowledge concerning energy — the topics, meanings, vague concepts and vaguer theories — there is a need to explore, if only in outline, the nature of general knowledge itself. Here, right at the beginning of the subject, is quite the toughest task of all. We have to find an answer to the question — is what we all know, or think we know — without ever having learnt it — a system of knowing at all?

The vast majority of our information about energy is not specialized knowledge of the kind learnt at school. It is a rag-bag of items which may refer back to episodes or conversation, family sayings, or to advertisements which 'caught the eye', and then almost inexplicably lodged in the memory. There are a number of possible reasons — personal, literary, perceptual, emotional or social — for holding fast to this knowledge, and the purpose of this chapter is to begin to unravel some of these.

General knowledge is certainly not the kind of trivial pursuit which figures in 'Brain of Britain' and other public competitions. That is a flagrant misuse of the term. What such games are really about is *non-general knowledge* because they aim to catch out competitors, eliminating all but those with exceptional powers of recall. Only in one way does such quiz-type general knowledge resemble truly socialized general knowledge. It consists of discrete pieces of information thought to be eminently reliable. Indeed, we are forced to rely upon them with a child-like faith because, unlike theoretical or systematic knowledge, such 'facts' do not come from a familiar field of knowledge.

Getting to Know about Energy in School and Society

'What is *quercus ilex*?'
'When was the Field of the Cloth of Gold?'
'What are the main constituents of cement powder?'

To a small body of specialists these items are known because they are a part of the system of things that they have learnt to understand. The answers have a place in an ordered scheme of knowing. To the rest of us they are little more than cognitive clutter, valued more for their rarity, and for the kudos that their knowing brings, than for any real interest.

True general knowledge is general because, we assume, everyone else also knows it too. But a moment's reflection is enough to conclude that they could not possibly all do so. Even if we could list every item of general knowledge about energy known to any one person, the tally would be bound to be different for the very next individual. For this kind of knowledge it will be the *assumption of similarity* rather than its actual existence which will be the important characteristic.

Talking and Arguing

There is a reason for this grossly improbable assumption of similarity. General knowledge is valuable because it contains *meanings* rather more than facts. We can only talk to other individuals if we believe that the words we use and the allusions we make will be comprehensible to them too. We need continual reassurance that they 'know what I mean?'. Explanation over the telephone, where the usual visual clues to comprehension — nodding or smiling — cannot operate, is particularly unnerving because we are deprived of reassurance. Social beings that we are, we depend on continual affirmation in order to continue with our spoken train of thought. So, just to be able to talk to other people, we must assume that we share general knowledge meanings. As G.H. Mead put it, 'we need to exchange perspectives' with the people we are talking to.

Conversation has always been the vehicle for common meanings and associations. Generally it carries on without effort because we all have an investment in the kind of agreement which is very little different from simply understanding. Since words have several meanings depending on context (a problem area which will be discussed in greater detail through the talking and writing of children) all intelligent listening is an active 'hopping' game as we move from one meaning vantage-point to another trying to find a place from which all the landscape makes sense.

Nevertheless arguments do take place in such conversations whenever we do not get ready confirmation and want to change the other's meaning, or lack of it, to our own. For this we use rhetoric — which has had a bad press over the last many centuries. We tend to use the term pejoratively now to describe political tub-thumping and illogical verbal bullying, but in this case we need the word in its basic sense of common, alogical persuasion. The ancients contrasted logic and rhetoric with care since they practised and valued both. Zeno, we are told, compared logic and rhetoric to 'the closed fist and the open hand'. It is a little surprising to find that rhetoric is considered the more open, and logic, which mathematics and science have appropriated and elevated to the higher position, as closed. But the reason for this openness is the two-way personal persuasion which is involved. To convince another by rhetoric we need to 'show' them how we are thinking. For this we need to work hard at producing exemplar material in situations which they too will recognize. So, in the language of modern sociology, we need to 'construct' how the other is, and choose those extra pieces of general knowledge which we feel will be familiar to them.

Rhetoric is not illogical: it just avoids logic. It is altogether looser than logic since it allows the essential change of context which the act of commonplace persuasion needs. The 'rhetorical question', which is almost all the substance that most people retain of this ancient discipline, is not just a question which needs no answer. It is the result of an open process in which the persuader seeks for a context in which the other would see things from the same perspective. Then, if successful, the punch-line in rhetoric, no different in this respect from the conclusion in a logical Socratic argument, simply confirms the end-game by showing that there can only be one answer. The underlying skill in rhetoric, but not in logic, is construing the other person's likely meaning and point of view. In logic, on the other hand, the character and knowledge of the adversary is of no interest whatsoever.

> The context of rhetoric is marked by justification and criticism, logoi and anti-logoi. It is a social concern in which different points of view clash, and there is a potential infinity of these clashes. The maxim of Protagoras (*that there are two sides to every question*) suggests that an unarguable rightness and wrongness cannot be established since critical challenges are always possible. Matters are different in the realm of logic...Deducing that 'Socrates is mortal' from the premises 'All men are mortal', and 'Socrates is a man' does not involve entering into an endless argument between religious believers and sceptics about immortality. (Billig, 1987:95)

So the consensual meanings of general knowledge are built up by almost identical social processes whether there is initial agreement or disagreement.

The general knowledge of children and adults is not likely to be quite the same because they have had different learning experiences. Nevertheless parents tend to assume that what they know is also the ultimate goal of what their children should know. It seems to be the aim of every community to pass on to the next generation, by means both open and devious, the knowledge that members of that community believe they share. It is a kind of cultural immortality in which education takes on the role of Dawkins' 'selfish gene'. General knowledge is the very stuff of common culture.

So adults talk carefully to children, as anyone who has heard a friend 'change gear' when a child enters the conversation, will bear witness. It is as if we mete out our stock of knowledge little by little, as and when we consider the child is ready for it. In the words of Joshua Meyrowitz:

> Socialization can be thought of as a process of gradual, or staggered exposure to social information. Children are slowly walked up the staircase of adult information, one step at a time.

However it is also true, as the same author points out, that the media now exposes children as never before to the full uncensored thrust of adult culture and information. Not just sex and violence, which raises so much alarm in some quarters, but adult musings on every aspect of life, have become totally accessible.

In summary, adult general knowledge about energy is likely to be insidious and persistent. It is continually reinforced by the meanings used in common talk, both person to person and through watching similar interaction on the television screen. In this way it is shot through with the kind of social understandings that we use more for understanding people than for scientific concepts.

The Operations of Common Sense

Mainstream general knowledge is what Schutz and Luckmann (1973) have called the 'social stock of knowledge' — a choice of words which carefully refrains from suggesting that it has either structure or internal logic. It might be no more than a random collection without any system — just titbits of information that have come our way in different circumstances. But human beings have to justify what they know, however it has been derived. The stock of knowledge may not be logically coherent; parts of it, as we shall see, will not stand up to even the most tolerant inspection. Yet it needs some kind of credentials — psychological if not logical — if we are to use it, and go on doing so. In place of a philosophy our collection of facts and opinions has a loose justificatory system which is defiantly referred to as 'just common sense'.

To say that something is 'good common sense' is to appeal to a silent but supposedly assenting majority, like those we imagine having confirming conversations with, and thus to dodge all argument. It also refers the matter to the arbitration of elusive principles which are thought to be enormously convincing, and yet are never named. In this way it can become a hiding place for both outrageous prejudice and total ignorance.

In his book on *World Hypotheses* Stephen Pepper defined three traits of common-sense knowing. The first of these is best described as 'taken for granted'. If it were otherwise, if the general knowledge we justify as common sense could be analyzed and confirmed in a more logical fashion, then it would cease to rely on the disreputable arguments of 'common sense' and become instead part of another and more formal and logical knowledge system, such as science. Common sense has to allow for the inclusion of values and feelings as well as facts, so the normal processes of logical analysis cannot accommodate it. Yet, because it is such comfortable 'taken for granted' knowledge, it does not even notice this absence of logical justification.

Pepper's second and related property of common sense is its *security*. To abandon, even for a moment, what is held to be common sense is like entering a looking-glass world where everything is unfamiliar and horribly disturbing. Learning school science can be just like this. It leads into regions where 'correct' principles seem to be grossly at odds with common sense. Who could believe, for example, that a bouncing glass marble is reversibly 'squashed' on impact like a soft rubber ball? We are taught in science lessons that potential energy *must* be stored in this way *in order that* the logical coherence of another knowledge system be maintained. Because children are still intellectually biddable they mostly go along, for a while at least, with this protection of the logical bastions of science; but it is at the expense of the comfort and security of common sense.

So scientific knowledge may require a kind of courage, even recklessness, which can be hard to keep up around the clock. School pupils often retreat back into the familiar arms of common sense when the tough logic of physics becomes too hard to maintain.

> No cognition can sink lower than common sense, for when we completely give up trying to know anything, then is precisely when we know things in the common-sense way. (Pepper, 1942)

The third characteristic of common sense is peculiarly awkward. Not only is it taken-for-granted, it can also flout almost every rule of argument and not be found out. Its security and obviousness seem to resent inspection. Social experiments which flout this convention by questioning everything which appears to be common sense, are frankly infuriating. Ask someone *in what way* they are 'feeling fine', or *why they think* someone 'looks shifty', and the greater your show of interest the more irritated they will become. The attitude of common sense is so alogical that it can tolerate neither attack nor probe.

When science writers like Joseph Bronowski recommended the thought of science as *The Common Sense of Science* because it 'asserts the unity of knowledge' we must treat the phrase with caution. The only unity in common-sense knowledge is the comfortable assumption of sharing. And that very comfort may depend on a belief that if we share the knowledge with everyone then no one will challenge it. In the words of Stephen Pepper common-sense knowledge is 'cognitively irritable'; like an amoeba it may even change shape when poked by an inquisitive outsider.

The fourth and final characteristic of common-sense knowledge is that it is *intersubjective*. For whatever reason it seems so obvious and secure to us, we expect it to be so also for others. That is not to claim that this vague knowledge-attitude provides us with similar cultural views about energy because they are based on similar observations and deductions. They are similar because they have achieved intersubjectivity — the result of deliberately matching our accounts to the views of others by watching for signs of understanding and agreement. Common-sense knowledge is born and bred in our everyday chat; it is reinforced and added to there, and achieves its greatest polemic triumphs there in the shifting ground of social arguments. These arguments are won not by logical confrontation, but by the art and practice of rhetoric.

All these curious characteristics of general knowledge will be important for studying children's ideas about energy. In particular they give strong hints about the way in which we should go about the task of exploration. Knowledge which is 'cognitively irritable' might be expected to squirm and even transmute under the probing of an intensive interview. Indeed, there is no shortage of published transcripts which show just such retreat and reinvention under the pressure of continued questioning, however gently it is carried out. On the other hand a knowledge system which is believed to be held in common, would be expected to be most itself in a social setting. Many of the illustrations to be quoted will be extracts from children or adults talking to each other in groups.

Personal Knowing

The characteristics of common-sense thinking which protect it from either inspection or introspection also differentiate it sharply from another kind of untutored knowledge. Some children and some adults have built up their own idiosyncratic theories and hypotheses about the world out of active private reflection. Such homespun explanations of the way things are can be described, tested and argued about. This is valuable 'personal knowledge' and not unreflective common-sense knowledge.

The boy Tim, for example, in Rosalind Driver's book *Pupil as Scientist?*, held a private theory that gravity was greater higher up because objects released from greater heights fall faster. He set out to test it by raising the arm of the clampstand to find out if the extension of a spring was greater at a greater height. This kind of personal knowledge which may even invite disconfirmation through experiment, lies poles apart from elusive common-sense knowing. It is just as much a kind of proto-science as that of the young

Einstein who imagined running after a beam of light, and later deduced that the speed of light is invariant.

The conclusions of personal knowledge, like those of common sense, may be either 'right' or 'wrong' in the light of the prevailing judgements of science, but they differ totally in their methods of justification. Personal proto-scientific notions are the products of private work on an idea or image, teasing and testing it to find how far it will match reality. It will be strongly championed in conversation. Elusive common sense is not open to such examination: it needs to remain flexible and even indefinite to match the meanings of those who may be engaged in conversation. It has little pretension to logical coherence. Phrases like 'Well it all depends on the circumstances' can be used to ward off attacks of inconsistency. As Schutz and Luckmann wrote, such life-world knowledge has a small-scale 'horizon of meaning'. Common sense has traded in logic in return for popular agreement. It is what everyone knows, and no one really wants examined. But personal knowing is private, treasured, and defensible.

Talk, Meanings and Historical Relics

There are at least three different ways in which we might go about the task of finding the common meanings for 'energy'. We might simply reach for the dictionary, especially for one which conveniently lists the different accepted meanings along with their first dated use in the literature. This is easy to do and we shall begin our investigations here.

In the second place we could listen to groups of people talking together about energy and try to infer their meanings, contexts and attitudes. We shall expect rather opaque and general purpose opening gambits as they begin to search for points of contact. It is where one person answers another that the agreement on meaning might begin to emerge. Recordings of the discussions of three different groups of adults who knew each other well, will be used for this purpose.

The third way of exploring meanings is more arcane. We will assume that at least some of these meanings might be relics of past theories about the nature of energy, albeit somewhat damaged and disfigured, as relics so often are. This supposition is based on the stability of metaphor or figurative speech in the language. There are so many examples of out-of-date terms, ranging from 'harnessing energy' to 'reaping benefit', in which old ideas are submerged in modern usage. It seems likely that words like these exercise some influence on meanings even if their original familiar impact has been lost. So it is profitable to refer to ideas in the history of energy to identify nuances of meaning in colloquial speech about energy. This will also be done.

Each method has its own problems. Dictionaries are far too sharp in their authoritative meanings. Discussions may be better, but only if we resist the temptation to interrupt and ask for explanations. Because of the nature of the common-sense attitude, such requests rarely bring any satisfaction. Sometimes the participants give a non-response or a repetition; on other occasions they seem to feel rebuked, fall silent, or search their memories for something which might sound more scientific and impressive. This does damage to the flow of taken-for-granted talk. The third method could provide an endless historical study in itself. It is the use of all three methods *in combination* which probably provides the best chance of sketching in some of the outlines of the meanings of energy in the general population. Three general themes were identified by this exploration.

I An Immaterial Agency

In classical times the word 'energy' derived from the world-view of Aristotle and meant something like the *potentiality for change*. Physics itself was derived from the Greek word for change so that this meaning gave energy a central position in scientific thought. However, even by the standards of Greek philosophy where immaterial essences were considered to be more profound than perceived reality, energy was an essentially immaterial affair. By the time the English word is first recorded in literature this essentially abstract quality still seems to linger on.

> They are not effective of anything, nor leave no work behind them, but are energies only. (Bacon, 1606)

> The light of the Sonne is the energie of the Sonne and every phantasm of the mind is the energie of the soul. (More, 1642)

> ...powers and energies that we feel in our minds. (Bentley, 1742)

The common phrase 'pure energy' to denote something at a very far remove from the imperfections of matter, shows that this ancient view of energy still lingers on.

It is sometimes suggested that physics is a perversely abstract activity whilst common thought is not. This is quoted by those who want to excuse the shortcomings of pupils, and is a part of the educational polemic which seeks to find a kind of science which may be more accessible to all. The objective may be laudable and faultless, but the argument — that abstract ideas are foreign to everyday thought — does not stand up to examination through conversation.

All the talk about energy recorded for this exploration was full of abstract notions. None of the adults had ever studied physics and yet, soon after the discussion was set going by asking what was meant by the word

General Knowledge about Energy

'energy', there were explicit attempts to emphasize the abstract nature of energy.

> 'A motivating force...which can include fuels but is also something more abstract.'

> 'Energy is a state of mind, it's how you feel.'

> 'At the heart of every living cell there is pure energy.'

> 'The sun is the symbol for energy.'

It is very easy to find other contemporary evidence for the abstract, almost anti-material, sense of the word. Popular science fiction is riddled with this attribute of energy. A creature of 'pure energy' is one without a material body. Teletransportation, a commonplace of television space sagas, is said to be accomplished by changing people into 'massless energy' (a somewhat worrying phrase for those of us familiar with Einstein's mass/energy equation!). 'Fields of energy' are used to transfix the merely material, and occasionally superior races of beings acquire the 'mental control of energy'. All of this might be said to constitute fragmentary and rather bizarre evidence that the most modern of popular space fiction perpetuates seventeenth century meanings of energy, which themselves may well derive from a metaphysical concept of Aristotle's which dates back some two thousand years earlier still.

II The Living Force

There is another general meaning attached to the word 'energy' which has been powerful at different times in the development of science. This is the connection between *energy and life*. By the seventeenth century it existed in non-scientific literature in a way which is not unfamiliar to a modern reader.

> Energie is the operation, efflux, or activity of any being. (More, 1642)

> When animated by elocution it acquires greater life and energy. (Holder, 1696)

By the eighteenth century phrases such as 'animal energy' or 'life energy' became much more common. The strict etymological sequence of the word is rather hard to disentangle, partly because these scientific theories originated in languages other than English, and also because the words used in this early science had not yet been fixed so tightly by formal definition as they are today. Well into the nineteenth century the words 'force', 'energy' and 'power' were still interchangeable. But the connection between energy — or some similar word — and life, was explicit. By the early nineteenth century it was to become the focus of a scientific controversy which will be described in chapter 9.

The history of the modern concept of energy might be taken to begin in the seventeenth century with the works of Galileo, Huygens, Newton and Leibniz. Each one of these founding fathers of modern physics identified a mathematical concept involving matter and velocity, or power, which seemed to have special and enduring significance. Of course there were high level arguments about the value of each formulation but, like most mathematical statements, they probably had minimal impact upon the general knowledge.

The works of the German mathematician and philosopher Leibniz are difficult for a modern lay reader. As co-inventor of the mathematical calculus there was bitter rivalry between him and Isaac Newton. While most of his books are punishingly hard to understand, his theories of force and movement include a curious strand of what may be the common thought of his times. He introduced the notion of creation, and of life itself, into the dry calculations of mechanics, and by so doing may have made contact with a strand of meaning already existing in the general culture. He identified a mathematical entity, '*vis viva*' which he calculated by multiplying the mass of a body by the square of its velocity — mv^2, and this he held to be the link between religion and mathematics. As an aristocrat and a scholar, Leibniz would probably have denied any influence from cultural general knowledge with great indignation. However, from the perspective of another age, it is easier for us to glimpse the meanings of a religious world-view beneath his properly academic protestations.

> But if the law decreed by God at the creation left some traces of itself impressed on things...then this indwelling force is one of those things which are not to be grasped by the imagination, but only by the intellect. (1670)

Although others before Leibniz had pointed out the continuing, quality of motion and mass in some mathematical combination or other, he believed that he had defined a quality which partook of the God-like nature of creation. It was not only a mathematical description but also an *internal function* of the object. No wonder then that Leibniz was to name this in-dwelling force '*vis viva*' — the living force. Because of its simple relation to mass and motion it applied to all moving objects and carried the implication that they were all alive. Stationary objects might have a 'solicitation to motion' but, because they had no actual velocity, they could only possess *'vis mortuus'* — dead force. (This, it seems, was rather like the inertia or mass of an object.)

> A substance is a being, capable of action...consequently all nature is full of life.

Leibniz's views may also carry some responsibility for the eighteenth theory of Vitalism with its quasi-scientific notion of 'vital force'. Unlike *vis viva* this force was not quantifiable, but it shared with it some other important characteristics. Vital force was also indwelling and enduring, being handed on like an olympic torch from live creatures to their living progeny. Indeed, vitalists were to claim that it was this vital force which had kept the species alive since the time of God's Creation, and was responsible for the energeticness and health of each surviving individual.

On the other hand Leibniz may just have been responding to an age-old animistic notion already present in the general attitude. In a realm so vague and chaotic as general knowledge, disentanglement of the cause and effect of ideas is a hopeless task. At all events his term *vis viva* was widely used in scientific circles, in place of 'energy', for more than a century. In the way of language, whether metaphorical or terminalogical, it was almost bound to influence how his contemporaries thought about energy.

Returning to the groups of modern adults discussing their ideas of energy recorded in the 1980s, these earlier vitalistic notions of energy provide another way of responding to their talk and meanings.

> 'Energy is something you are enthusiastic about and that gives you the push you need.'

> 'I always think of energy as being a re-vitalizing thing.'

> 'All the created energy....'

III The Measurer of Work

It was only in the nineteenth century, when the great steam engines of the Industrial Revolution had already thundered away for more than a century, that the notion of mechanical power became impressed upon the cultural imagination. Now the meaning of energy finally acquired its modern dimension of work and power. This time there was no haphazard picking up of an old meaning out of an existing basket; the word 'energy' was deliberately introduced by Thomas Young in 1807 for a new purpose. On the surface he did no more than redefine *vis viva* as simple 'energy', but in this vigorously industrial age it came to imply work and power, rather than any mystical indwelling animism.

> The term energy may be applied with great propriety to the product of mass or weight of the body into the square of the number expressing its velocity.... Some have considered it as the true measure of the quantity of motion; but although it has been universally rejected [in favour of momentum] yet the force thus estimated well deserves a distinct denomination. (Young, 1807)

> In almost all cases of forces employed in practical mechanics, the labour expended in producing any motion is proportional, not to the momentum, but to the energy which is obtained. (Young, 1845)

In the thirty-eight years which elapsed between the first and second of these quotations Thomas Young had come to see the quantity $(½) mv^2$ as related to real everyday work and hence of more practical importance than mv — the momentum that Newton had used in his laws of motion. Such had been the national prestige of Newton and the blind chauvinism of the British scientists, that any suggestion that Leibniz's rival notion of *vis viva* could be of even equal value with momentum was 'universally rejected', as Young said, for over a century.

Names for concepts, like the meanings of words, do not change instantly at the command of a mere physicist. *Vis viva* did not die immediately; 'force' continued to be used by British scientists in contexts where the modern school child would be severely reprimanded for so doing. Only by the latter half of the century did 'energy' come to be conceptually linked with 'work'. This meant that it could be measured, even when the object was not moving, by the work equivalent transferred (e.g. raised water or a charged battery). The next step was to try to freeze this conceptual definition through the text-books of physics.

But scientific text-books probably had even less influence on general knowledge in the the nineteenth century than they do today. Science was not a school subject. For those lucky enough to have secondary schooling at all, classics was usually the exclusive diet. However, the nineteenth century was a prime time for the popularization of science. There were scientific societies in big cities like London and Manchester, and scientific journalism was beginning to flourish. While the 'Lakeland poets' were trying to make their writing accessible to the 'common man', Humphrey Davy, amateur poet and intimate friend of Coleridge, was trying to do the same for science through his lectures and experiments at the Royal Institution.

The following extract is taken from a nineteenth century issue of the journal *Nature*, which was founded in those more liberal times to spread interest in science, and has now become one of the many scientific journals whose dry and specialized articles are read only by the academic.

General Knowledge about Energy

> We live in a world of work from which we cannot possibly escape...the energy which a labouring man possesses means, in the strictly physical sense, the number of units of work which he is capable of accomplishing. (Balfour Stewart, 1870)

Then, in the manner of any good teacher trying to put an unfamiliar idea across, the author draws analogies with more familiar notions. He chooses a social and hierarchical comparison which reads as though it would indeed be readily accessible to a readership of upper middle-class Victorians.

> Energy in the social world is well understood. When a man pursues his course, undaunted by opposition, he is said to be a very energetic man.
>
> By his energy is meant the power he possesses of overcoming obstacles; and the amount of this energy is measured (in the loose way we measure such things) by the amount of obstacles which he can overcome — the amount of work which he can do. Such a man may in truth be regarded as a social cannon-ball.
>
> By means of this energy of character he will scatter the ranks of his opponents and demolish their ramparts. Nevertheless, a man of this kind will sometimes be defeated by an opponent who does not possess a tithe of his personal energy. Now why is this?...If two men throw stones at one another, one of whom stands at the top of a house and the other at the bottom, the man at the top of the house has evidently the advantage.
>
> So in a like manner, if two men of equal personal energy contend together, the one who has the highest social position has the best chance of succeeding. For this high position means energy under another form. It means that at some remote period a vast amount of personal energy was expended in raising the family into this high position....

It is not difficult to see a psychological objective behind this writing, over and beyond the pedagogic. Balfour Stewart is deliberately taking down the forbidding precision of physics. He uses a frankly outrageous comparison between the measurement of energy and unquantifiable personal enterprise — 'in the loose way we measure such things'. In this, and by his use of simple and even racy language, Balfour Stewart makes it possible for anyone to talk easily about energy. Common-sense habits would shy away from measurement and hard definition. Only when the meanings of a concept become flexible enough to be used metaphorically can they enter the realm of general knowledge and so become culturally important. Within the modern group of adults there was some connection between energy and work in their talk. Only two participants used the actual word.

'Energy is what you need to make things work.'

Another said:

'...the energy needed to power factories.'

This is not conclusive evidence that energy is rarely seen by the majority of people as something that 'works' or 'powers' machinery. In this small group of conversations, however, it was not nearly as common a meaning as were many of the others. The measurement of energy in work units had hardly entered the common stock of knowledge. In some groups, where the talk centred on 'mental energy', it was even denied that energy was the kind of concept which ever could be measured.

The popularization of science has now acquired a medium far more pervasive than journals. Television is present in 98 per cent of British homes. None of the adults taking part in the discussions had learnt formal physics at school but one of them displayed a remarkable range of knowledge which he claimed to have gained entirely from watching science programmes on television. He spoke not only about nuclear power and pumped storage hydroelectric power stations, he also mentioned matter, anti-matter and the creation of energy out of mass. It is a characteristic of television science that it attempts to impress, astonish and entertain by using spectacular examples. This individual's contributions to the discussion included the same non-mechanical social notions as the others, but he also made distinctly different points which included the measurement of energy.

> ...energy in the forms of waves and wind, because you can tell from the power of these things that there's energy there, no matter how it is brought about.... You can calculate what is going to happen as a result of the release of energy — how much water there is and how far it is going to fall.

Metaphors and Sigmund Freud

Words commonly change their meaning by becoming swallowed up in clichés, comparisons, and metaphors. This may begin by a hesitant comparison: one of the discussants said, 'Energy is rather like strength of character.' This is a well-distanced simile which compares one thing to another while keeping their basic meanings well apart. And yet no comparisons can be made without effecting some changes in the meanings of each. In this simile 'energy' is used far more in the sense of stored work in a spring, than for the movement of a machine: so the kinetic meaning of the word goes into abeyance. 'Strength of character' also has its meaning altered. It becomes more like the will to do work and less like, say, the moral rectitude needed to stand by a friend who is wrongly accused. Similes and metaphors are just as interactive as they are comparative.

This process by which a metaphor, if often used, begins to change the meanings of words, has been examined by Max Black (1962) and others. Black rejected the common view that metaphors are no more than comparisons or substitutions. He wrote that:

> In the simplest formulation when we use a metaphor we have two thoughts of different things *active* together and supported by a single word phrase whose meaning becomes the resultant of their *interaction*. (my emphasis) (p. 38)

If such a metaphor is used often enough it permanently reorganizes the meanings that we have for each word on its own, and then the figure of speech gains familiarity but ceases to cause surprise or reflection. It has simply filled a gap in the common meanings that are available to us.

The result of a word being used in a metaphor so frequently that it becomes a cliché, is that the metaphor is 'dead'. It is no longer perceived as a comparison at all. Both 'falling in love' and 'dropping asleep' must once have been fresh and arresting. Now defunct as metaphors, they survive as ordinary modes of speech. In the same way 'harnessing hydroelectric power' now conjures up no images of horse tackle: it is simply the appropriate term for a process for which there is no other. The complex of energy terms and meanings has become permanently changed and enriched.

The term 'mental energy', which seems as though it might have been coined by just such a metaphor, has a more interesting history. The first definition of energy broad enough to embrace all its manifestations and to include conservation, was made by Helmholtz and a like-minded group of young reductionist physiologists in the 1840s (*see* page 181). One of these friends was Brüke who later became Freud's teacher. When Freud turned from medicine to psychology he had a similar reductionist ambition; he wanted to make his work as similar to the mechanistic physics of the times as he possibly could. That would give it far greater status.

Accordingly Freud began his manifesto, *Project for a Scientific Psychology* in 1895 with the sentence, 'The intention is to furnish a psychology that shall be a natural science.' To do this Freud imported many terms from mechanics including energy, work, excitation, force, and power, and in the beginning he tried to make them as precise as he could. About energy he wrote to a friend (Elkana, 1983) that it obeyed 'the law of the constant quantity of mental energy'. But despite Freud's intention this borrowing of the term energy only made it a hidden metaphor. During the course of his life he found more and more difficulty in using mental energy as a conserved quantity. Later he seems to have tried anew with 'libido', and failed again.

However, the public clearly finds the dead metaphor 'mental energy' very useful. The adults in conversations about energy used it almost more often than any other manifestation of energy.

General Knowledge Serving Personal Ends

People take enormous liberties with general knowledge meanings, because they are felt to be so comfortable and consensual. This means that the three different themes of meanings which have been described so far are not sharply differentiated in general conversation. Even for the handful of comments quoted above it is easy enough to find ones where both the notion of massless abstraction, and of living force, can be found together, or where power for work is combined with mental motivation. Amalgamation is one method by which the definition of energy is continually being changed and blurred by usage.

The most striking feature of this talk about energy was the way each of the different participants used this same word to describe their widely differing interests. One spoke almost exclusively about doing housework, another about non-violence and Eastern philosophy, several felt themselves prompted to speak about psychology, or the politics of oil pricing and consumption. For another it provoked an outburst against computers in education.

The operation of metaphors still does not quite explain how it is that one word may be used in such disparate personal ways. There is one further stage in the blurring of meaning which needs to take place before a word can achieve this extraordinary pliability. When a word is frequently use with a large number of different meanings it begins to be seen not only as useful, but as capable of stimulating a variety of powerful images all loosely related to one important theme. The cultural salience of this theme is crucial. If it sufficiently seizes the imagination the word becomes a 'symbol' — thematic and powerful within that culture. This is then followed by an explosion of popularity in which the word is directly invoked, without even the effort of making a metaphor, by anyone who feels that they can cash in on it. 'Atomic' has had that power since the bombing of Hiroshima, now 'environmental' is launching its bid for universality. 'Energy' has been symbolic for much longer than either of these.

Inevitably this process blunts the meaning of the word at the same time as sharpening its value as a universal image. This is close to the poetic use of words like 'dawn', and 'death', which are well understood to be symbols in either verse or newspaper headlines, indeed rather more often than their literal meanings. Poetry places great value on the vagueness of symbolism, in a way that science cannot do.

In the talks by our groups of adults the use of the term energy revealed more about the person's conception of the world than it did about the semantics of energy.

> 'I regard energy as warmth, comfort, light, personal light over my bed so that I can read.'

> (asked what energy means) 'I would have assumed you were just talking about oil; it's to do with politics and economics.'

> 'It's getting up in the morning. (Laughs) Something to go on and on and not having to have a breather...you know...the capacity to do more. Other times it's things you feel you have to do, you know, like the ironing which I hate but you've got to do it though, haven't you?'

> 'To me, when you mention the word, it brings to mind our western civilization. It seems to me that energy, being energetic, is characteristic of Western civilization. To me, I associate energy with violence, with aggression, thrusting...I associate it with a restless spirit.'

The idea of general knowledge as shared meanings has now been stretched almost to breaking point. What started as fluid but intersubjective meanings have become idiosyncratic and personal. Yet there is no real contradiction here. Social conversation serves to express individuality and even eccentricity, as well as cohesion and conformity.

The one supreme criterion of meaning is whether or not other people can agree, not so much to the sentiments expressed, as to the validity of its usage. They must find it comprehensible and viable. As one woman said after a particularly eccentric definition of energy followed immediately after her own more familiar idea:

> 'Well, it did not immediately occur to me, but as soon as he said it, I knew what he meant.'

Chapter 2

Schools of Thought on Children's Notions

Examining Different Ways of Thinking

In the last chapter there were few constraints on how to begin, or where to search. It was a new field. But as soon as the subject becomes children's ideas related to school learning we encounter whole teams of other researchers — educationalists, and both social and cognitive psychologists, even linguists — exploring the same territory. This implies that there will not only be more available knowledge, but also a variety of points of view.

There are many influences on children's learning, both internal and external, which are continually causing growth and change. Adults' ideas do not remain stationary either, but the pressures to change towards a more 'correct' or 'formal' goal are rarely so strong as they are for school children. This is one substantial reason why so many researchers are eager to study the area. They want to find out how changes come about and how they may best be promoted. It thus becomes both a government funded service industry enlisted to improve schooling, and also a convenient laboratory for the study of cognitive development.

Trying to find out what either children or adults think about familiar happenings is unlikely to be a value-free activity. Our own attitude towards children and towards learning are almost bound to intrude. In anthropology, for example, right up to the 1930s there was a distinctly patronizing attitude towards more primitive peoples' views of the natural world. Their witchcraft and totem rites were examined and reported as symptoms of inferior or defective mentality rather than as coherent belief systems in their own right. Such offensively superior attitudes have long since been abandoned by the anthropologists, but in the case of young children the situation is more complex. Any educationalist must believe that a child develops and matures intellectually. It would follow that their early thinking might be expected to be immature, and its conclusions faulty, at least by comparison with those of an adult.

Piaget's Contribution

The early work of Jean Piaget began the exploration of children's ideas. It was novelty enough to investigate the way that children thought, and to listen to their own faltering words, at a time when (in the first decades of this century) the educationalists were simply devising rote-learning rules for more efficient school learning. Their best offerings were the Law of Recency (keep on repeating it!), and the Law of Effect (reward them if they can repeat it correctly!). Piaget's two early books on young children were about their ways of thinking seen as a progression towards the logical reasoning of mathematics which Piaget clearly believed to be the pinnacle of all thinking.

Then, in the introduction to his next work, *The Child's Conception of the World*, Piaget wrote the following:

> The form and functioning of thought are manifested every time the child comes into contact with other children or with an adult and constitute a form of social behaviour, observable from without. The content, on the contrary, may or may not be apparent and varies with the child and the things of which it is speaking. (1929)

That distinction between the forms of thought and the content of ideas still stands today, although few people nowadays, I suspect, would go along with Piaget's order of difficulty. There is still the same division into questions about how children's thought operates, and about what views they hold, but the development of the interview technique which he pioneered has made the second question seem easier to tackle than the first. Literally thousands of children have been asked why they think things happen the way they do, and the results have proved to be of tremendous interest. Piaget's later work was almost all of the first kind. It outlined a stage theory of development in which different mental processes became available to the child, one after the other, as outside stimuli and internal mental growth released new capabilities for logical thought.

This elaborate system of 'genetic epistemology' has come under some attack. This is only partly due to conflicting experimental evidence. Another reason was the movement of educational rhetoric away from an external assessment approach, and towards a more child-centred one. The movement for exploring the content of children's own ideas coincided, more or less, with the arrival of mixed ability comprehensive schooling in Britain. If the achievements of all children were to be equally valued, an approach which focused on the children's ideas, rather than on the stage of their cognitive development, was likely to seem less elitist and more in keeping with the current educational ideology. By the same token it cannot be denied that if the Education Reform Act of 1989 manages to reverse this trend towards value-free exploration of children's own notions, as well it may, interest can be expected to revert to the discipline of stage development of cognitive growth which Piaget marked out. Neo-Piagetian research has not stagnated during this period. It has begun exploring social influences on logical growth within the framework of genetic epistemology (e.g. Perret-Clermont, 1980) and the possibilities of connecting success in school science with Piagetian stages (Shayer and Adey, 1981).

Piaget spoke to hundreds, if not thousands of children, but he presented the content of their answers in categories not of the situations encountered, but of the internal mental stages of development. One of the most memorable passages in his early book, *The Child's Conception of the World*, concerns ideas about the sun and the moon. These strange animistic notions are presented less for their own interest and more as an advance on an earlier mental stage where thought is believed to be connected with the voice or the air. Now, Piaget argues, there comes a stage at which children believe that objects can be magically manipulated. Here is one passage from an early interview that he quotes in his book:

'Does the moon move or not?'
'*It follows us.*'
'Why?'
'*When we go it goes.*'
'What makes it move?'
'*We do.*'
'How?'
'*When we walk. It goes by itself.*'

New Educational Research

Piaget was not an educationalist. He had already enumerated the philosophical, scientific and primitive ('savage') ways of explanation, and was now intent on investigating the explanatory methods of children by empirical methods. Yet for many teachers, the charm and indeed the impact of this discourse lay not in the epistemological arguments, but in the actual content of children's ideas. (If anyone doubts that the moon appears to move in this way they can try it for themselves by going out on a moon-lit night. It does not matter whether they drive, run, walk, or simply twist their heads: every movement made is mirrored by a movement of the moon. Perhaps they can no longer believe, with the simplicity of the young child, that it must be their movement which causes the motion of the moon, but they may be no better able to explain why this happens than the little seven year old in Piaget's interview.)

Passages like that throw up quite different questions from those which occupied Piaget. He went on to ask what would happen if someone else moved and in a different direction. Would the moon follow them too? The children tended to say, yes, they thought so. From this Piaget identified an ego-centricity and lack of logical consistency in the children's thinking. But when research into children's ideas began anew, during the 1970s, many people simply wanted to collect more of these naïve ideas, as might an anthropologist. The questions they were exploring did not concern logical cognitive development, but the differences between these ideas and those being taught at school. They wondered where the ideas originated. Did the children really construct their own considered explanations for natural phenomena, and test them against experience? — were they descriptions, explanations or misunderstandings?

Two factors were necessary for the explosion of research to begin. First, it had to attract the attention of those engaged in science education research whose own scientific background was sufficiently informal to entertain the possibility of alternative explanations. Scientists and science teachers have an unenviable reputation for unflinching adherence to the 'right answer'. The communal nature of the scientific enterprise, where the worldwide community of practising scientists publish their results in a uniform format, is a strong force for keeping all their explanations in step. One of the triggers for new educational research was a new approach to science itself which would release some of this rigidity. By the 1970s the new, more relativist philosophies of science had arrived. These made it entirely possible to contemplate and study multiple understandings of natural phenomena. Thomas Kuhn himself, one of the foremost in this movement, acknowledged a debt to Piaget in the preface to his seminal book *The Structure of Scientific Revolutions* (1962).

> A footnote encountered by chance led me to the experiments by which Jean Piaget has illuminated the various growing worlds of the child and the progress of transition from one to the next. (Kuhn:vii)

Secondly, it was necessary that the earlier mode of social science research, which had laid so much emphasis on reproduceable laboratory tests, should either pass away or be deliberately flouted. Listening to children talking, especially in the teaching laboratory, would need to be validated by quite different procedures from those of earlier experimental psychology. In practice, however, when in 1973 Rosalind Driver based her thesis on the recorded classroom comments of pupils, participant observer research had just become respectable. It was now possible to suspend both assessment and judgement in favour of listening and recording. Children's pre-scientific notions could be explored for their own interest.

Towards Alternative Frameworks

Like Driver in America, Erickson in Canada (1976) and Tiberghien and Delacote (1976) in France began with small-scale studies. These were about heat and temperature — simple everyday aspects of energy about which young children might have developed their own ideas. In the latter case it was just two twelve-year-old girls who spoke about their ideas on heat throughout the year in which they were supposed to be learning the official scientific version at school. Then the link with school classroom began to diminish. An early piece of research by Edith Guesne (1976), also in Paris, used drawings and recorded interviews to find out what children thought about light and seeing, without any reference at all to the school curriculum.

The results of these investigations were surprising and intriguing. One little French girl stuck obstinately to her belief that wrapping a thermometer in wool would be bound to raise its temperature — because 'wool is warm' — for the whole year in which she was being taught about heat in school. Others added two temperatures together to calculate the final temperature of the mixture. Many children believed that light travels from the eye to the object being 'looked at', the better to probe its shape and colour; or that light does not travel at all but just is there all around the inside of any well-lit room. Because the researchers were trying to understand the children's ideas in the manner of a stranger in a foreign culture, rather than being judgmental about the way of thinking behind them, these explanations took on their own rationality. They began to seem like a valid though unusual way of thinking — even though they were at odds with the accepted scientific paradigms.

By 1978 this kind of research was growing in popularity and there were at least two important landmarks in the field. One was a very substantial piece of research into school and university students' ideas on mechanics by Laurence Viennot. This masterly piece of work succeeded in probing the students' views, setting them in a scientific context, and even drawing attention to cultural and linguistic influences of the kind that were examined in chapter 1. Parts of this study were very influential, and are described in a later section.

There was also a thoughtful review article by Driver and Easley (1978) which performed at least two important tasks. In the first place they coined an apparently neutral term to apply to these ideas of children. They called them 'alternative frameworks', a name still used by many to describe this whole field of inquiry. Secondly, they made clear an important distinction between two different approaches to pupils' understanding of science. In the first place there was the 'nomothetic' attitude which compared the children's ideas to the taught theories of science. The use of such a yardstick makes a judgmental attitude towards the children's ideas almost inevitable.

On the other side was the neutral 'ideographic' approach, which aimed to give naturalistic studies of children's ideas, seen as the world-views of a different culture. Using the jargon of a Kuhnian philosophy of science these were 'paradigms' — ways of looking at phenomena similar in kind, although not in content, to those produced by science. 'Pupil paradigms' became a rhetorical term.

Or Another Science?

Then events started moving faster. In New Zealand and in England (Surrey) a new, more ideological, approach began to develop. In retrospect it is now clear that a complete divorce from school learning was almost inevitable if the research attitude was to be value-free. In New Zealand the transition was particularly abrupt and seemed to arise in the following way.

Many of the early researchers, like Beverly Bell and Miles Barker who explored what children thought an 'animal' was, had themselves been practising science teachers. Predictably, their work was related to school learning and often included the simple action research that a gifted teacher might put into effect in order to set matters right. Then, these teachers were joined by researchers with a less teacherly orientation. Some of them began to catch in their tape recordings, not only the pupils' ideas about phenomena, but also their reflections upon learning and authority.

> ...and the teacher's way is the right way...that's what I find hard...I would say to the teacher that he teaches the theory his way, what he says is right...and gets every kid in the class to write down their way of seeing that...not what he has written, but their way...(Osborne *et al.*, 1980)

This was a challenge to which this school of research was particularly vulnerable. Were alternative ideas really just alternatives, like jam or marmalade? Were they of equal value with the theories of science? Was there any real sense in which only the teacher's ideas were more scientific? Perhaps it was their anxiety to accord adequate respect to the pupils' views which made this challenge so powerful. The New Zealand school of research argued the matter out and finished by adopting the term 'Children's Science' for these untutored ideas. This was a deliberate polemic designed to emphasize the coherent and theoretical nature of children's explanations.

Very soon after this the Surrey school of educational research, which worked closely with the New Zealanders, began publishing papers which reinforced this perspective by referring to George Kelly's *Theory of Personal Constructs*.

> Each man contemplates in his own personal way the stream of events upon which he finds himself so swiftly borne.

The Kellian theme of 'every man his own scientist' began to be used in educational circles in much the same way as Paul Feyerabend's work was used in philosophical ones. Both seemed to deny anything special in the nature of scientific knowledge or method. Feyerabend's nihilistic comment 'anything goes' (1978) was much quoted in radical educational circles. Naïve theories of objective scientific 'truth' had long gone out of fashion. Middle of the road philosophers of science accepted that there was a limited kind of relativism in all scientific knowledge. The most superficial acquaintance with the history of science showed that theories which had seemed perfectly secure in the past were later toppled by new evidence. The same might happen to the accepted explanations of today.

Nevertheless most philosophers had continued to claim, with Popper and Lakatos, that scientific theories were constructed and tested by a special social and logical methodology. No one would claim that a young child could construct their explanation of phenomena in the same elaborate way; indeed most children would not claim this for their own ideas. There was a sense in which Feyerabend was an essential ally for the 'Children's Science' protagonists against the more staid modern philosophers. His work could be seen, at one remove, as offering a rebuke to the remaining adherents of Piagetian theory of cognitive growth. Why should the final stages of logical ability be required for having alternative scientific explanations?

The empirical work on children's ideas went on apace in most of the countries of Europe and in North America. There were investigations — all as neutral and value-free as could be — into children's notions about movement, force, energy, living organisms, electricity, evolution, gravity and many other scientific topics. (*See review by* Gilbert and Watts, 1983) But there were two characteristics of the research findings which did not fit comfortably into the 'Children's Science' movement. Liz Engels (Leeds, UK), Audrey Champayne (Pittsburg, USA), and many others, carried out research which did not confirm the picture of children as scientists logically applying their cherished theories in appropriate contexts. Whether it was the concept of pressure (Engels), or motion (Champayne), the children slipped cheerfully from one 'alternative framework' to another in the course of questioning in an interview, or over longer periods of time.

> The students...never notice that a proposition they have used to explain one of the situations is directly contradicted by the proposition they use to explain the motion in another situation. (Champayne *et al.*, 1980)

Those who had taken up ideological cudgels on behalf of 'Children's Science' were hard to discourage. Michael Watts, for example, published an article on 'Some alternative views of energy' (1983) in which he stated his position clearly:

> Such ideas and meanings for words are not simply isolated misconceptions, but are part of a complex structure which provides a sensible and coherent explanation of the world.

Yet, in the same article, Watts reports the following pupil explanation of electrical energy which hardly seems to be consistent about the direction of flow.

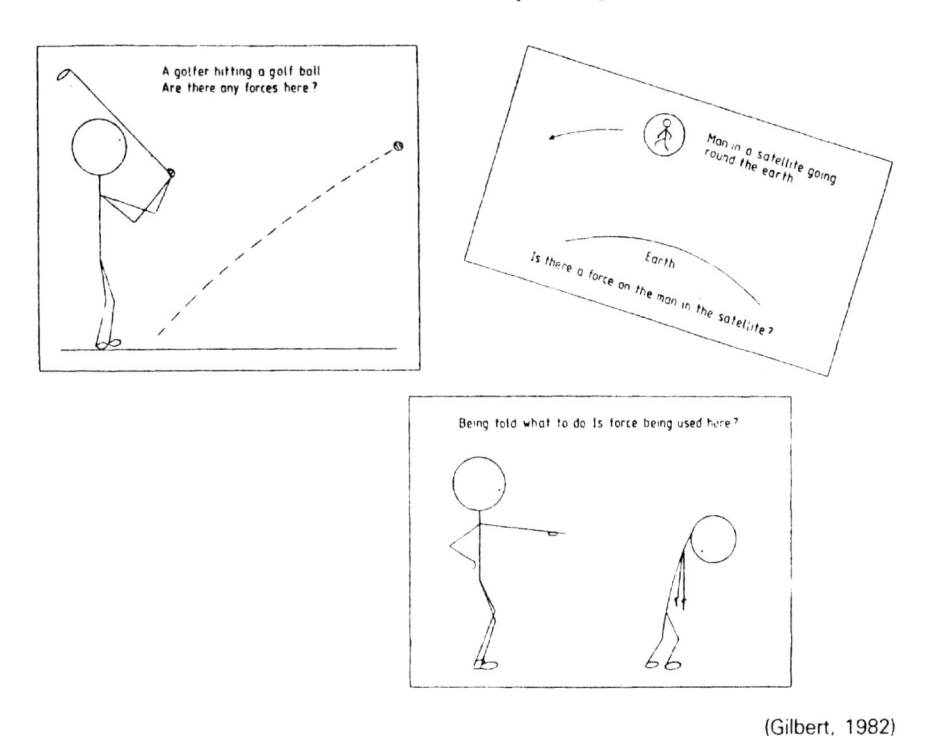

[Figure: Gilbert, 1982 — cartoons asking "A golfer hitting a golf ball. Are there any forces here?"; "Man in a satellite going round the earth. Is there a force on the man in the satellite?"; "Being told what to do is force being used here?"]

(Gilbert, 1982)

...*the energy comes out from both leads*...because you never get a circuit without the other one...*it comes out of the negative end*...flows round the circuit...encountering the light bulb on the way...where it can transfer some of the energy...*and goes back to the battery* (my emphasis) (Watts, 1983)

John Gilbert, also working at Surrey University, and others had explained that this represented an amalgam of 'Children's Science' and 'Teacher's Science' (1982). (The relativistic attitude towards knowledge insisted that neither system could be simply designated as 'science'.) The amalgam was shifty and inconsistent; yet still there was no ideological retreat. To these authors it remained both 'sensible' and 'coherent'.

The persistence and commonality of children's notions were curious in another way. Studies which aimed to teach (and so should more properly be left to the next section) showed that even when school science had been learnt with apparent success it still tended to succumb, with the passage of time, to some of these same original ideas. This suggested to some researchers that there might be a kind of insidious social reinforcement of the children's notions. This could cast yet more doubt on the idea of an individual and personally constructed 'science' knowledge.

This academic discussion may yet be preempted by political forces. A centrally controlled educational system dedicated to raising standards through nationwide controlled assessment cannot easily tolerate deviant ideas, even from children. It is easy enough to foresee that a more judgmental attitude will eventually supersede the respectful and *laissez-faire* alternative frameworks movement. Indeed, the recent work of Shayer and Adey (1990) may even constitute an early confirmation of a trend of this sort.

Misconceptions — They Got it Wrong!

This second school of research differed fundamentally in the attitude it took towards the children's knowledge. If light were the topic, for example, and many of the pupils appeared to believe that light went out of their eyes to probe the object being looked at — rather like a blind man with a stick — this was not to be hailed as an interesting neo-Platonic theory constructed by the child to explain reality. However careful or consistent the children's explanations, from this perspective the idea was judged to be basically incorrect. Furthermore, if the children had already been taught the right explanation — that of school science then it was worse still — a mistake had been made.

Informal research into the common mistakes that children make in school must, in some sense, be as old as the first school classroom. Any teacher sighing over a pile of homework where a large group of pupils make the same mistake, will probably plan to start the next lesson with some remedial action. For hundreds of science teachers, it was their daily experience (or was it research-on-the-job?) which taught them what misconceptions to expect. Teachers have always known, for example, that children expect there to be less electric current flowing in a circuit after the light bulb than before it. Year after year their pupils have claimed that this must be so because 'lights use up electricity'. Any teacher with a spark of originality then plans new teaching designed to correct this fault.

For researchers of this persuasion it went without saying that the prime purpose of schools, and the educational research which serviced them, was the better learning of 'correct' knowledge. This kind of research was judgmental, school-directed, and 'mission-orientated'.

Research on Learning

Laurence Viennot's work on force and motion displayed many of the well-known school mistakes — like identifying the force with the direction and magnitude of the object's velocity. More interestingly, she demonstrated that such mistakes took place more readily in familiar everyday problems, than in abstract ones. This led her to consider *not* that these ideas were a

Schools of Thought on Children's Notions

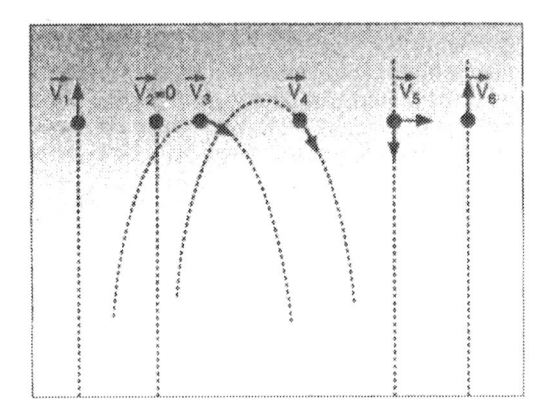

result of consistent personal reflection, but that they were produced in everyday unreflective exchanges. She began as follows:

> The work described here has its origins in practical teaching problems, and its ultimate aim is to contribute to the improvement of teaching.... We shall show that such spontaneous reasoning constitutes not just a few mistakes made by some students, but a way of thinking found in everyday conversation and in much that one reads. (Viennot, *op. cit.*)

One of the problems Viennot presented was 'a set of juggler's balls, captured in flight, all at the same time.'

Then she asked another parallel question which, in the typically abstract way of physics, was set in a hypothetical world. 'If the same force acts on two identical masses are their motions necessarily identical?' This time the results showed a *higher rate of success* than in the pictorial question. This seemed at first to be a curious result. Should it not have been the case that thinking about everyday events was easier than thinking about abstract ones? It was not an isolated effect.

About the same time there was a substantial doctoral thesis by Leon Pines at Cornell University in which he interviewed a large number of elementary school pupils about seeds, living, electricity, and energy. (Much of the information he obtained on the last of these has interesting points of similarity with some of the general knowledge about energy established in the last chapter.) Then he devised a teaching scheme, on cassette, in order to correct these concepts which were, in his own words, 'riddled with misconceptions'. He found, alas, that they were also very hard to move.

Pines' conclusion was that the children's notions were far from being a coherent, quasi-scientific scheme. He reported that they 'change over time, are complex, multifaceted...highly situational and context dependent.'

Doctoral students are expected to systemize their data and divide it into neat illuminating categories, but this proved to be extraordinarily difficult with the wealth of vivid material he had collected. In frustration Pines wrote:

> It is true that specific responses can be categorized but almost any child will exhibit responses characteristic of many categories, irrespective of the category system used.

From the perspective of general knowledge this result would not have been so hard to comprehend. Meanings in the general culture have to be accessible to everyone; these children had already, even by the age of seven or eight, heard enough of the varied uses of these words to possess a range of meanings for them. But in the 1970s university departments of education were looking for tidier systems of thinking.

The third example comes from Scotland. This was a large-scale investigation undertaken on behalf of the Scottish Department of Education by Mary Simpson and Brian Arnold (1982) to look into the mistakes made in the O-grade biology examination which was taken at age 16 by the more able students. Much of the work was statistical but, because it was imaginatively combined with pupil interviews and reflections by their teachers, it proved far more illuminating than such official studies usually are.

The first topic was respiration and it produced overall results that were similar to those of Viennot. If the questions were about everyday terms — about food, energy, or growth — the answers were often wrong. If they were about more technical and unfamiliar biological concepts such as gas transfer across membranes they were actually more likely to be right. The second topic for investigation was photosynthesis and their results are delightfully summarized by the cartoon illustrated opposite.

Conceptual Change

For some educationalists of this 'misconceptions' persuasion, arguments based on the current philosophies of science proved just as attractive as they had been from the perspective of personally constructed science. From America and South Africa came papers which took up the idea that acquiring new knowledge in the classroom might be like the problems of establishing a new theory in the world of science. They argued that for children to learn school science something like a 'thought revolution' might be required. Suddenly, as at the time of Copernicus' assumption that the earth was moving round the sun, all the familiar happenings had to be seen in a new light. It was what Thomas Kuhn had referred to as a 'paradigm shift'. If all students' observations were coloured by the theories that they held, then 'conceptual change' to a system based on totally new concepts and meanings

Schools of Thought on Children's Notions

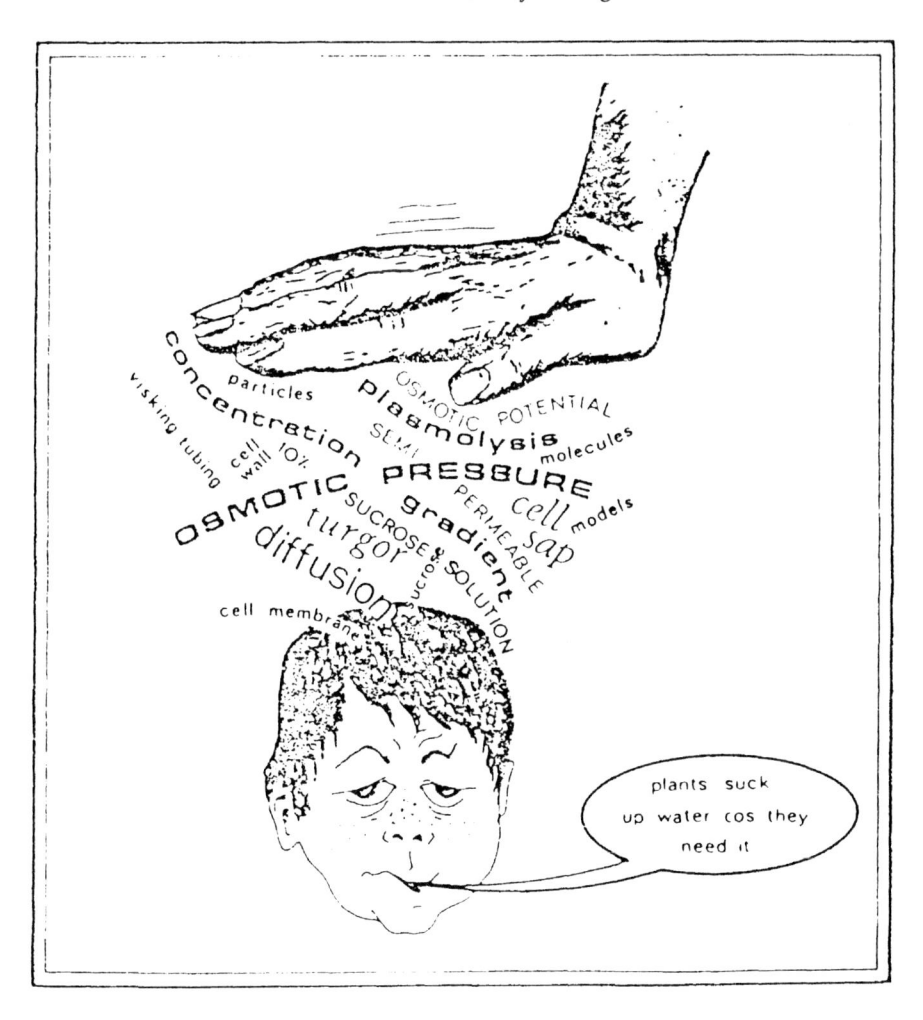

might be hard to achieve, especially if the context continued to remind them of more familiar ways of knowing.

There was a brief flurry of theoretical interest in studies of how people change their opinions. In particular these theorists argued about producing an essential 'cognitive dissonance' which might lead to change (Festinger, 1950). If enough conflicting evidence could be produced, they claimed, pupils would be forced to change their opinions to avoid the mental discomfort of trying to accommodate a falsified prediction. For a science teacher raised in the empirical discovery tradition of the 1960s this idea was very appealing. The 'voice of nature' as it were, revealed through unequivocal experimental results, was all that was needed. Cognitive change would follow. Thoughtful educationalists at Cornell University drew up a list of the necessary requirements — a recipe for making cognitive change. There had to be dissatisfaction with the existing idea and an alternative explanation which was more satisfactory. Once again these authors argued that the learner was like a scientist, and so supremely rational. They wrote:

> ...rationality has to do with changing one's mind...
> Instruction needs to focus...on building a rational bridge from current concepts to new ones. (Strike and Posner, 1982)

This eminently sensible approach was developed in the logical minds of educational scholars. At heart, however, it was contrary both to the nature of common-sense justification as we examined it in the first chapter and also to the notion of Kuhnian paradigms. The latter suggested that the children's observations and interpretation of experiments would be coloured by the notions (paradigms) that they already held, so contradictions would be very hard to see. But even if they were observed, common sense, as we have seen in chapter 1, can go its own unthinking way by simply ignoring all the rules of logic.

Nevertheless, showing up the contradictions between what the pupils' misconceptions would predict and actual events did seem to be a possible strategy. It tackled the real problem, ignored by so many of the convinced Kellians described in the previous section, of how to teach in the teeth of these alternative notions. Research students, however, found it hard to put into action in the classroom (Smith and Lott, 1983).

Putting it into Practice

The 'alternative frameworks' group of researchers had done little more than recommend that children be allowed to discuss their own ideas together. In this 'misconceptions' school of research the professional objective was to improve the learning of correct science. A number of recipes for cognitive change and programmes of action research designed to effect it were reported. A few appeared to bring some success, but it was often only transient. Verdicts were different; whether or not they could report eventual success, all these researchers were agreed on one point — these children's naïve ideas were obstinately persistent.

Quite the largest piece of action research so far undertaken by teachers is the CLIS (Children's Learning In Science) Project. Well funded, staffed by enthusiastic teachers and a team of experienced researchers from Leeds University, and with a clear run of five years at their disposal, this team set out to show how teaching may start from the children's alternative frameworks and yet end up at the finishing post of accepted scientific explanation. They took up three special areas of science — plant nutrition, particle theory of matter, and energy.

It is probably too soon to give a verdict on this work although the accounts of experiments which the children, or the teachers, designed to confirm or disconfirm competing theories showed some of the same characteristics that others had found. Once again the children's ideas were hard to shift.

> There was a tendency for some groups to gloss over anomalous results in order to seek reinforcement of their original hypothesis. Some pupils may report that the experiment is 'done wrong' because some unexpected results are obtained. 'I believe there is some protein in soil and the experiment that said there was no protein was done wrong.' (CLIS)

What has stood out most prominently in the published accounts of the CLIS project was neither simple cognitive change, nor a sure-fire scheme of teaching. It was that discussing ideas, their own or those of science, fascinated many of the children. For most science educationalists this came as a considerable surprise.

Persistence and Theory

It is tempting to sustain a good story-line by claiming that these results themselves initiated a linguistic approach to the exploration of children's pre-scientific notions. But this was not the only possible explanation. It was agreed by many that the persistence of children's naïve ideas might well arise from 'theory-laden' observation, that we see what our imaginations have prepared us to see. So far this only begged the question of where the naïve theory came from in the first place. There were three suggested answers.

Andrea diSessa (1981), supposed a perceptual origin to these naïve ideas. He called them 'primitive notions' which are able to stand 'without significant explanatory substructure or justification'. These phenomenological notions, *p-prims*, are recognized in terms of what people perceive in the world around them. While it is not clear if this adequately explains the genesis of such ideas, their patchy distribution and multiplicity, it is not hard to see that such p-prims would be persistent, as our results require, if all perceptions of happenings tended to reinforce them.

Since structuralism was enjoying a very high public profile at this time, it is not surprising to find the suggestion that, like Chomsky's linguistic structures, these naïve ideas might be genetically determined. In fact there was nothing so very modern about the basic idea. Kant had described a priori structures of the human mind, and Preece examined this possibility in the context of children's misconceptions.

The Kantian hypothesis — that many of the templates by which we structure the world are innate rather than constructed — provides a parsimonious explanation of much that is known about children's science. (Preece, 1984)

Language and Experience

The linguistic explanation was the third possibility for children's pre-scientific notions. The resistance and commonality of children's notions had suggested to several researchers that the root of these ideas might be found in the culture itself, and that the ideas got transmitted insidiously by the language. A little research along these lines had already been carried out. The necessary supporting linguistic theories were in place and the argument had been made (Sutton, 1980), but it was just not being used.

It is possible to draw a strong connection between the language of a culture group and the perceptions of its members. There is an often-repeated observation that Eskimos have about twenty different words to describe the stuff we commonly name 'snow'. Now it is clear enough that their chilly and continual experience of snow might well give rise to all these different terms. Everyday experience must affect the development of language, since it sets the scene which language needs to describe. That is a fairly trivial conclusion. But supposing I, who am not an Eskimo, have the greatest difficulty in observing, or even of imagining, just five different types of snow. Does the absence of words for the other variations deny me the capacity to distinguish between them? Could language control my very perception?

The argument about language, knowledge and perception has occupied sociologists, psychologists, and linguists for many decades. G.H. Mead had begun from a sociological perspective in which the communication of ideas through language was responsible for creating at least a part of the experience. He wrote that 'the social process of communication is responsible for the appearance of new objects in the field of experience'. Mead was trying to extricate attention from the metaphysical notion of an 'inner experience' — probably primary sense perception — which was not itself communicable, and so could not be shared with others.

Children's talk, like that of adults, is the stuff of communication and its role in constructing experience appealed directly to the teachers and researchers examining children's communications. In a book written by a group of gifted teachers of English the personal effect of communicating experience was eloquently described.

> We all need to work through, sort, organize and evaluate the events of our daily lives. Sleeping we do this in dreams, waking in internal monologues and relaxed talk. As individuals we have to assimilate

our experiences and build them into our continuing picture of the world.... So we hold out to others, in talk, our observations, discoveries, reflections, opinions, attitudes and values, and the responses we receive profoundly affect both the world picture we are creating and our view of ourselves. (Martin *et al.*, 1976:16)

Most of that point of view fits very easily into the perspective of general knowledge in the first chapter. The concluding words are new; this affective aspect opens a new dimension to understanding.

The Sapir-Whorf Hypothesis

The connection between language and perception could be even tighter than the sociologists had claimed. In the 1940s the linguists Benjamin Whorf and Edward Sapir, put forward an hypothesis strong enough to provoke both attack and experiment. In its most uncompromising form it asserted that what we believe about the world around us — even what we perceive and experience — is *actually dictated by the language* that we have available to express it.

> ...observers are not led by the same evidence to the same picture of the universe unless their linguistic backgrounds are the same.
>
> It is the grammatical background of our mother-tongue, which includes not only our way of constructing propositions but the way we dissect nature and break up the flux of experience into objects and entities to construct propositions about.... (Whorf, 1956)

Whorf was quite aware that his hypothesis, in its strong form, challenged the whole perceptual basis of the sciences. The first to react were psychologists. Experiments were set in hand to find out whether primitive peoples who had no word for some particular colour — it might be turquoise — could actually distinguish and experience it as well as those who did have a special word for it. It seemed that they could. Similarly the deaf mute were just as good at recognizing colours as the hearing who knew their names. But it was not difficult to show that memory for colours relied heavily upon using their familiar name-words as the address systems for recall. The deaf mute did far worse on such memory tests.

If this form of the Sapir-Whorf hypothesis did not stand up to testing, a weaker form might do so. Whorf himself, and later the sociologist M. Halliday (1978), proposed a formulation in which it was not so much that we could not perceive states for which we did not have a word, but that the words we have and use serve to direct us towards a particular point of view and way of thinking. If language affects what features of a situation children pay attention to, even if it does not actually distort their powers of perception, then it sounded like a theory which could well be important for science education. It could provide a viable answer to the 'persistence of alternative notions' problem.

Just what this means for an understanding of concept words, such as energy, which do not seem to rely directly upon sense perception, can be illustrated by some bilingual anecdotes. It is well known that young English-speaking children hold a concept of 'feeling energetic' which they easily confuse with energy (*see next chapter*). The work of the German educational researcher Duit (1981) showed that for young German children the commonest association for energy is electric current. This does not mean that Germans never suffer from listlessness, or that they do not notice when they feel tired without cause, as the strong form of Whorf's hypothesis might suggest. Instead, when going to the doctor to complain of what we might call a 'lack of energy', the Germans speak of a lack of *Kraft*. This word is commonly translated as 'force', not 'energy'. Just how difficult this whole area of translation can be was highlighted by an episode when I mentioned this in a seminar on the meanings of energy. At the end a young man came forward to correct me; he had been brought up in Germany until the age of 7, he said, and was sure that the word *Kraft* meant 'energy'. At that age, it seemed, the relevant associations for energy and *Kraft* were so similar that they directed thinking towards the same experiences. Neither the concept of electric current, nor of efficiency of transformation are matters of any great importance for most seven year olds.

Some further conversations and light-hearted comparisons with German speakers suggested the following table of meanings. The connections between words and experiences were different in the two languages, but neither language-culture suffered from a lack of vocabulary in which to express them.

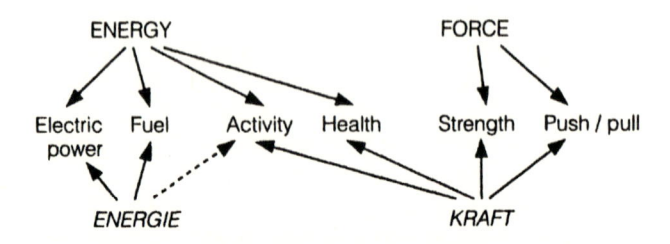

ENERGY → Electric power, Fuel, Activity, Health

FORCE → Strength, Push / pull

ENERGIE → Electric power, Fuel, Activity

KRAFT → Activity, Strength, Push / pull

Cross-cultural Educational Studies

One of the first pieces of research designed to test the effect of language on children's understanding of science concentrated on the concept of speed. It compared four- to five-year-old Japanese children, who have the same word for 'early' and 'speed', with Thai children who have different words. They were all shown two cars running across a table. Car A starts first and runs slowly. Car B starts later from further forward, runs faster, and reaches the edge of the table only just behind car A. Most of the Japanese children said that car A travelled faster than car B, presumably because the meanings of the word used made them concentrate on the earliest car to reach the table edge, rather than just on the speeds of the cars. The Thai children did significantly better on the same question (Mori *et al.*, 1976).

In the field of energy Duit followed up his study of German-speaking children with a comparison of their associations with those of Philippino pupils (Duit and Talisayon, 1981). There was a clear difference. The Third World children rarely gave either power stations or electricity as one of their first reactions to energy. The difference was clear enough, but if this work is to be interpreted on a socio-linguistic basis, rather than just in terms of common objects, more information is required about the mother-tongue of the children concerned.

A more illuminating piece of research (Ross and Sutton, 1982) was undertaken with secondary school students in England and in the Tiv-speaking region of Nigeria. This enabled the researcher to examine the effects of mother-tongue and school language on the development of scientific concepts such as growth, heat, energy and electricity. The method used was to collect the word associations of Tiv pupils educated in Tiv, Tiv pupils educated in English, and English pupils educated in English. The results showed that most of the same associations were present for all three groups of children, but in different proportions. Both groups with Tiv as their mother tongue gave similar numbers of the different associations.

In the case of 'light', for example, Tiv speakers gave more meanings involving:

lamps and electricity, and
helping us to see.

The English students referred more often to

similarity with brightness, and
a form of energy.

The connection between lamps and electricity is interesting. We learn that the Nigerians use metaphors such as 'light of the fire' for electricity, as well as the term *latriki* which is itself a corruption of the English word. So it came about, the authors reported, that 'electricity was a special form of light for many of the Nigerians'.

If spontaneous thought and its associations were carried on in the language of childhood, and if this language directed attention, emotions, and memory, then these were just the kind of results to be expected. It would not be at all surprising if Nigerian children visualized the flow through wires in an electricial circuit as more like light through optical fibres than the common notion of current. This is not so much an individual view of phenomena built up by personal perception and reflection, as culturally affected imagery impressed by the language used.

The word association approach to children's ideas need not be cross-cultural. Studies of how any pupil's scientific concept grows and changes can be carried out in a similar way. But if such tests yield no more than lists of associated words the result would be uninterestingly bitty. No one doubts that these clusters of words have some kind of local meaning structure which holds them together. It might be a developed piece of personal theorizing, a favourite imagined simile, or just a 'sedimentation' of related items which experience has taught us to expect together. Word associations on their own yield little information about how the meanings behind these clusters have developed.

Answers and Questions

The past dozen years of educational research has produced some results. They are listed below but look, in some respects, more like answers to questions never asked, than questions posed for a new research programme.

1 **Children do already have ideas about some of the phenomena that they will study in school science.**
   This includes everyday topics such as light, breathing, substances and energy.
2 **Their ideas have a great deal of commonality within one culture.**
   Language and social interaction are implicated.
3 **They are resistant to change, sometimes inconsistent, but almost always consensual.**
   Most of this evidence throws the argument back to the functions of general knowledge.
4 **Children find their own and others' ideas interesting to discuss and think about.**
   This is a new and valuable finding.
5 **Scientific knowledge sometimes acts as though it were easier to reproduce than everyday knowledge.**
   Formal knowledge may be learnt and retained in a different way from general knowledge.

In the field of cognitive science and artificial intelligence researchers of the decade were asking hard questions and seeking answers. In educational research no consensus between the three schools of thought — personal alternative frameworks, school misconceptions with the need for conceptual change, and socio-linguistic meanings in one cultural group — had emerged, so the theoretical questions were difficult to formulate. Like the progress of science itself, without some agreement about structures and paradigms educational research is in danger of doing no more than repeat earlier studies. Indeed, it seems that in almost every topic area there have already been several trawls of children's notions, delivering almost identical results.

On the Construction of Constructivism

The nearest the research community has come to a common perspective is through the use of the term *constructivism*. It was thought to be based upon the findings above but, like the research itself, is now interpreted in different ways. Originally the term was used to indicate that science was not given to us directly by the verdict of experiments but that it was 'constructed' by scientists. Different cultural groups might well construct different sciences. For some members of the alternative frameworks school, constructivism means that the pupils are not 'empty vessels' to be filled with school knowledge as one might top up a jug of water. There is no emptiness: children have pre-constructed ideas. That meaning itself has been constructed.

To others this body of research proves in some new way that pupils are 'active learners' and that this too is constructivism. It could be doubted if the work has really proved this, or indeed that it ever needed proving (Millar, 1990). Minds cannot have knowledge impressed passively upon them as one might brand a sheep. Learners have always needed to work very hard. When the teaching has not helped with this process, as in the worst of university lecturing, student learning proved to be very tough going indeed.

(It is true that very young children, and the cowed and frightened adult, sometimes learn unthinking behaviour — like not touching fire or instant obedience to a bully's commands — through stimulus and response. It is also the case that rote learning, word by word, will not necessarily bring conceptual understanding. Obviously the learning, however active, has been misdirected towards words rather than ideas. But real understanding, by definition, must involve active mental effort.)

Yet more definitions of constructivisms have become built into almost all the different research traditions.

> Constructive alternativism suggests that people understand themselves and their surroundings, and anticipate future eventualities, by constructing tentative models. They then evaluate these against personal criteria, so as successfully to predict and control events based upon the model. (Pope and Watts, 1988)

Unfortunately evidence from research does not support this conclusion either. If only it did such 'scientific' pupils would have no trouble at all in learning the models of science by the same logical process. Alas, they do have trouble! And it may be precisely the sharp contrast between the two methods of knowledge construction, everyday and scientific, which is to blame. (This is examined further in chapter 6).

Unanswered Questions

Some hard empirical questions have been thrown up.

> How should teaching take account of these naïve ideas?
> What effect, if any, does age-related development have on the nature of alternative explanations, or on the ease with which they are changed or by-passed?

The first question has not been ignored by research, but it has not been fully answered. About the second question there has been much less research. The CLIS project has shown that different notions (but are they alternative frameworks or misconceptions of what they think they have been taught at school?) arise at different ages.

There is also some research on children's use of simile to describe electric current, which throws a little light on the second question (Solomon et al., 1987). It showed that over a two-year period pupils made a significant improvement in their ability to compare the idea of electric current with 'a river', or with 'fire', even though their general knowledge of electricity remained much the same. The younger pupils (age 11) often failed to make a proper comparison; they were more likely to write 'electricity is not like a river' (the given simile) *because* 'it is dangerous to mix electricity with water'. The capacity to construct metaphors, or to compare one mental model with another, plays an important part in forming new scientific explanations (Hesse, 1966). These data suggest that the conscious use of a model to illustrate an explanation of happenings may be age- or ability-related. Close similes comparing the look or feel of one thing with another are much easier. Perhaps we may conclude that the younger children's alternative notions may be descriptive similes more often than they are distanced and over-arching models (Solomon, 1986).

Martin Monk (1990) has claimed that the alternative frameworks of children can be examined from the point of view of consistency to show that they are related to Piagetian cognitive progression. Children's naïve ideas must be related to their increasing experience of the world. What they do with that experience inside their heads, may also change because of their capabilities to compare and infer. Piaget himself was convinced, right from the start, that children's own constructs illustrated their stage of mental development as well as their feelings.

Schools of Thought on Children's Notions

'We are faced' wrote the young Piaget, 'by an original tendency (to form constructs), characteristic of child mentality, penetrating deep into the emotional and intellectual life of the child.' (1929)

This is only the third place in this long chapter where the emotional reactions of the pupils have been evoked. Further discussion of these reactions will be found in chapter 8.

Chapter 3

Children's Ideas on Energy

Trigger and Response

In the following chapters a great range of evidence about the children's notions on energy, both before and during their school learning, will be presented and discussed. The data were collected over a three-year period, by the author, in a variety of ways so that the many aspects of the understanding of energy — private and social, cognitive and affective, linguistic and imaginative — could all be sampled.

How we ask the children for their ideas about energy depends upon our own research perspective. If we believe they are private and personally constructed ideas we shall choose to speak to children on their own, with confidentiality. If we think that ideas are socially constructed and fragile without the feedback of others' reinforcement, then we may look for a group of children so that they could argue it out together. Those researchers who want to trigger the word associations of children, or their imaginations might get them to list words, draw, act, or write a poem. The same child might well respond differently in each of these modes because clues in the manner of asking and in the social context give them different impressions. The gathering of information could seem either like a sociable occasion or a school assessment. Only by using several methods can the whole picture of children's understanding be properly explored.

Most of the data about children's ideas for this chapter will come from spoken or written words in answer to very general instructions given in the classroom. We begin with two poems written by secondary school pupils in Malaŵi who were in the middle of a course on energy. The special advantage of asking for poems is that this allows a much freer atmosphere than is normal in science lessons, one where individual expression and originality is encouraged. These two poems contain all the complexity arising from a mixture of taught science, African culture and individual metaphor, put together unselfconsciously without any attempt on the part of the child authors to make the different strands agree. They also show that wish to touch on the broadest and most universal themes which, as has been shown in chapter 1, a symbolic term like energy allows.

> I am invisible.
> I can be known by actions.
> Without me there will be no life.
>
> My name is energy,
> If there is no energy,
> there will be no life.
>
> The whole world depends on me.
> If there is no energy
> there is no work.
>
> If plants would have no energy
> animals would not exist.
> Without energy, there is no life.
>
> My dear brothers and sisters,
> I am a needy one
> I make things, or work, to be done.
>
> When I am manufactured,
> I am taken and stored,
> In people and animals.
>
> When people lift a stone
> I am transferred with a stone
> and force it to move
> and kill a bird.
> But, my dear brothers and sisters,
> I am not consumed.

Prose writing about energy may have less charm but it is probably easier to interpret.

'Write about Energy'

The next collection of data come from pupils in the First, Second, and Third Years of two English secondary schools who were given blank sheets of paper and simply asked 'What is energy?' This took place before they had been taught anything about the topic in school. Half the children attended a mixed comprehensive in north London; the others came from a semi-rural comprehensive, also mixed, in Yorkshire. After this first question, the pupils were then told to write three sentences showing how they would use the word 'energy'.

These were the simple triggers which launced a hail of scribbled ideas, associations and meanings. These responses, like the African children's poems, also led us into the world of their interests, imaginings, and ambitions. Energy, it seemed, could be located somewhere within everyone's favourite pastime. Thus the sentences gave hints about the character of the child as much as a selection of meanings.

'I need a lot of energy when I train for football.'

'Seb Coe has a lot of energy.'

'Old people do not have much energy so they need medicines.'

'The monster threw an energy bolt!'

These children knew no physics definition of energy, such as 'the capacity to do work', or 'what is transferred between systems when work is performed'. They responded with their own kinds of meaning for the word. In some cases it was clear that there was a conscious effort to explain their meaning, as well as to show it. There is an important metric in the different possible methods of explanation (Solomon, 1986). This is the 'explanatory distance' between the meaning of the word as the children see it, and the explanation or comparison they produce to show what they have in mind. In one mode explanation can stay very near the subject and be only a little different from the search for a *synonym*.

'Energy means power' was a popular response.
'Energy is strength' was another.

Explanation can also move into an over-arching position from which it may survey and compare several distinct contexts. Only from such a standpoint is it possible to make a *generalization* which might cover all uses of the word. Such responses were, predictably, far rarer than those of the first type and their discussion will be left to the final section of this chapter.

A Meaning Cluster

Starting with the first, less ambitious, type of explanation a simple count of the words given in answer to the trigger 'what is energy?' showed their relative popularity. When two or more words were given together the

Children's Ideas on Energy

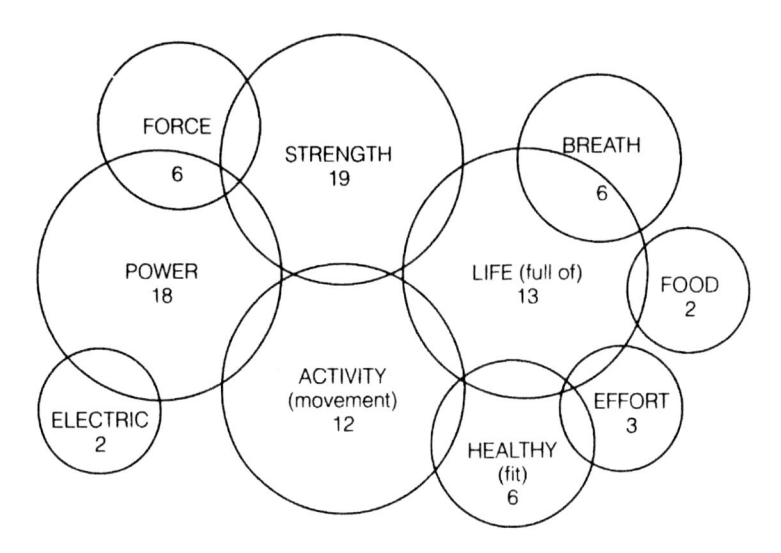

```
        FORCE
          6      STRENGTH      BREATH
                    19            6

      POWER           LIFE (full of)    FOOD
       18                 13              2

                   ACTIVITY
   ELECTRIC       (movement)       EFFORT
      2               12              3
                               HEALTHY
                                (fit)
                                  6
```

evidence yielded a little extra evidence about the possible grouping of words within a total meaning cluster. Using these clues a spatial map of the word cluster was devised in which over-laps showed where pairing occurred. The figure above shows this early attempt to chart the associated meanings of the words that first year pupils in both schools gave in answer to the question 'What is energy?'. (At this age there were only a small number of attempts at more generalized explanations — those of the second type — and these are not included.)

The rough organization of the cluster shows a trend from the non-human terms, like 'force' and 'electric' and, possibly, 'power', on the left to the more intimate attributes of people on the right. 'Activity or movement', like being 'healthy or fit' are not very surprising ingredients of the energy picture. The presence of 'effort' on the far right of the cluster may be similar to the talk about effort and mental energy recorded in the first chapter.

The most unexpected feature of this cluster was the high score on 'life' and the related term 'breath'. Sometimes these two words were used in tandem. The expression 'full of life', or more charmingly 'full of life and joy' may be metaphorical descriptions for 'being energetic', which seemed to be the underlying sense of many of these sentences. What was more surprising was the number of children, both in this year group and the next, who wrote simply, 'Energy is the source of life'.

This odd phrase sounds significant. Paper answers can be irritatingly terse and allow of no cross-questioning, so they need other references to suggest meaning. The out-dated notion 'vital force' which figured in chapter 1 may be useful once again. In the eighteenth century this force distinguished the living from the non-living and, like the elixir of life, was thought to be passed on from an organism to its off-spring. In this sense vital force was a kind of 'life source'.

It is not only in times past that it was considered entirely rational to suppose that movement and activity are indications of a life force. It is still a strong and unthinking association today amongst children. The reasons for this do not have to rest upon a kind of historical recapitulation where the young live through all the bygone theories of their race in progression before they can make progress towards new ideas. Vital force could well be a lingering association with a past and defunct idea, which has been passed on via common metaphor and ways of speaking, but it will also have characteristics of its own appropriate to today, and to children. One other difference to be expected between historical scientific theory and apparently similar contemporary ideas, is that the latter will be less structured and may apply inconsistently. Since it is a part of common-sense thinking it will be much harder for the children to explain and for us to interpret. Word associations will not be likely to be sufficient for proper interpretation, but they will form a starting point.

The Energy of People

Immediately after the pupils had made their attempt to 'explain' energy, they were asked to write three sentences to show how they would use the word. This may have been much easier to do. Some pupils actually wrote more than three, although others dried up after only one or two. This opening move in the research project delivered nearly a thousand different sentences.

The first task, confronted by such a richness of data, was to identify some very simple characteristic and to count its appearance or non-appearance. A great number of the sentences were about human beings, and a few mentioned animals. These were combined under the general heading *living*. Another large block made equally clear references to the *non-living* such as machines or wind. This first sort through left only a handful like 'The whole world needs energy', which were ambiguous and needed separate consideration.

Some of these sentences were egocentric. Many simply said 'I need energy when I run', others appeared to be slightly more detached, 'When you run out of energy you are out of breath', but there was little doubt that the reference was to personal experience. Piaget labelled one of the earliest phases of explanation as egocentric — like the moon following the individual child (chapter 1). As might be expected this type of sentence became progressively less frequent in the older year groups. In research on children's views on electricity the same effect was also noticed (Solomon *et al.*, 1987).

Children as old as 11 or 12 clearly do not believe that the phenomena of nature are still directed at them, as the younger ones may do. Nevertheless,

Children's Ideas on Energy

when they are asked to think about a new topic many begin from their own experiences. The most intimate and immediate of these learning experiences are those which come through the sensations and movements of the body. Motor memories may then imprint concepts in the mind. It is commonplace that our muscles seem to 'know' how to swim, or how to ride a bicycle, long after we believe in our minds that we have forgotten. When asked to write about energy it seemed that may of the children's most immediate reactions were related to their bodily feelings of exuberant fitness or of painful exhaustion.

Physicists often claim that these primitive bodily reactions to phenomena should be good guides for the young students of physics. Since physics attempts to explain common natural phenomena, movements and forces, there should be no difficulty in reconciling what children have derived for themselves in order to move around the world, and the theories that scientists construct to explain dynamics. The children may not have sophisticated theories but they will have correct rules of thumb which 'work', as indeed they must for moving around safely. These will guide both action and thought. Such intuitive understandings about the world have been dubbed the 'ultimate explanatory elements' by Goethe, and 'phenomenological primitives', or p-prims for short, by diSessa (*see* page 31).

Yet these p-prims do not prove to be so very reliable for learning concepts in practice, or in theory. DiSessa's own work showed, for example, that if the children were invited to play a computer game in which they controlled a moving 'sprite' which had to kick a goal, they often got it wrong. They only launched the 'kick' when the sprite was already moving across the mouth of the goal; and then shot directly at it instead of allowing for the sprite's velocity. They could aim quite successfully in reality when it was they who were moving and kicking the ball, but this 'muscle knowledge' seemed to be so localized that it could not even be used by another set of finger muscles controlling a computer figure.

It may be that, far from being a basis from which physics may be built, these p-prims are themselves made more accessible through learning the abstract knowledge of physics. The diagram on page 46 was used by McCloskey (1983) in a rather similar study of older students' mechanics. He asked them when the runner should release a ball for it to drop into the hole. Those who had studied physics got the answer right significantly more often than those who had not. Assuming that both groups would be equally good at dropping balls into a hole, this is a surprising result. Perhaps muscle knowledge does not easily translate into explanatory p-prims which can be verbalized or recalled so as to prove useful.

Of Girls' and Boys' Responses

Many of the children in the energy study referred to the effects of energy on their own muscles. The younger ones particularly spoke of sweat, and of

Getting to Know about Energy in School and Society

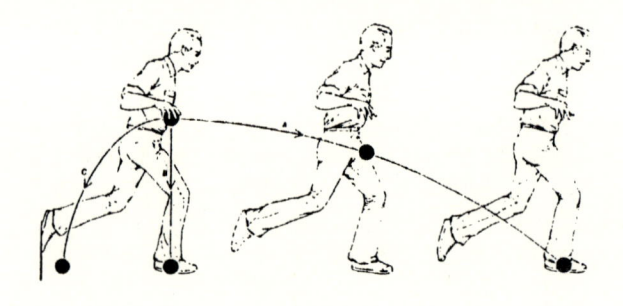

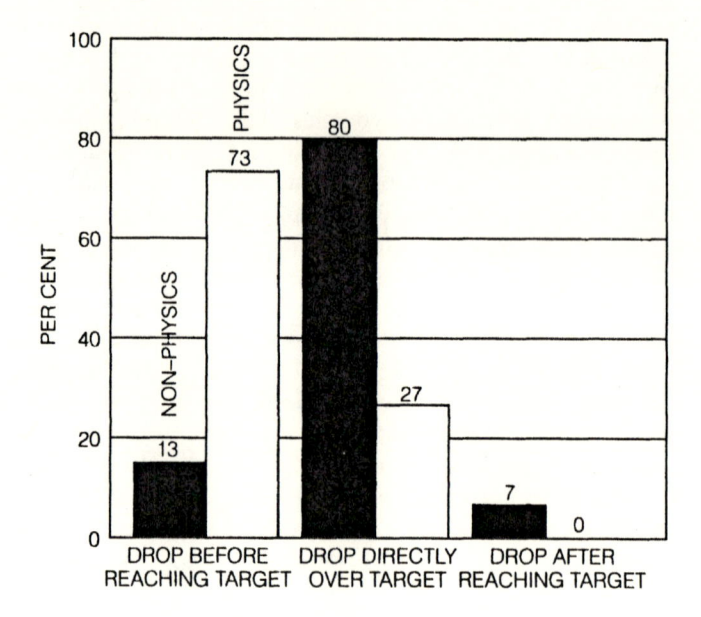

being out of breath, as being related to the using up of energy. Often they were confused and somewhat incoherent about these manifestations, and, as we shall see, their ideas did not always serve them well in terms of learning the accepted scientific concept. The p-prim concerned with feeling energetic, if such it was, may have come about through a connection between a meaning of 'energy' in everyday talk, and an experience of fitness in their own bodies.

Of course, living associations need not be egocentric. Some of the sentences were about the welfare and health of other people, or of animals. Both old people and young toddlers figured in many of the sentences, especially those written by the girls. In every group, First Year to Third Year, able or less able, the girls gave more sentences with living associations than did the boys. (In Table 1 the discrepancies are due to sentences which

Table 1 Statements about Energy

Year		Living (%)	Non-living (%)	Total
FIRST	Girls	93	6	174
	Boys	72	25	149
SECOND	Girls	70	24	194
	Boys	52	38	143
THIRD	Girls	50	42	135
	Boys	35	52	119

cannot be attributed unequivocally to either category.) For at least the last ten years the science education community has been anxiously trying to discover why it is that girls show so little inclination to study the physical sciences. More recent and careful testing show that it is not due to a lack of intelligence or of spatial ability, as had been assumed in some quarters. In all the research work that has been done to try to grasp a point of relevant distinction between the preferences of girls and boys — whether by statistical analysis of large-scale written responses, or by gentle long-term counselling and interviews — the result is almost always the same. Girls seem to be more *interested in people* than are boys (Collins and Smithers, 1984; Gilligan, 1983). Within the broad category of living associations another gender difference gradually became apparent.

Two Themes of 'Living' Energy

The pupils' sentences focused on different aspects of living creatures and their relation to energy — food or sport, a feeling of vitality or recollected exhaustion — and they gave evidence of quite different ideas about the meaning of energy even within this category.

Some sentences were too short to indicate much — 'We all (or all people) have energy', or they stuck so closely to a particular sporting interest that the conceptual meaning proved insubstantial. Eventually, however, patient sorting and re-reading were rewarded by indications of deeper and distinct meaning themes which, although they sometimes overlapped, showed that there were indeed different semantic usages. It was rare, however, to find a single sentence that differentiated clearly on its own. The scheme on page 48 is presented in an order which seems best to show the mounting difference between the two clearest meaning themes about human energy.

The left-hand column of statements defines a meaning for energy verging towards that of *health* and perhaps, a modern version of the life-force itself. Those on the right-hand side might be identifying a kind of *human kinetic*

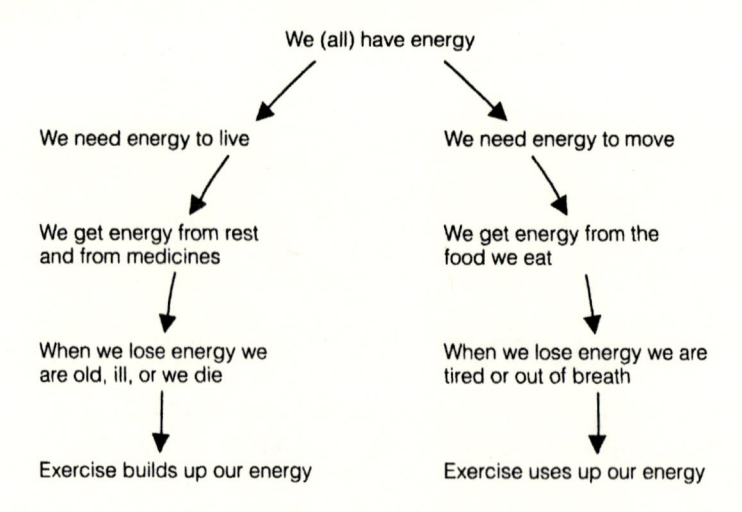

```
                        We (all) have energy
                       ↙              ↘
     We need energy to live        We need energy to move
              ↓                           ↓
     We get energy from rest       We get energy from the
     and from medicines            food we eat
              ↓                           ↓
     When we lose energy we        When we lose energy we are
     are old, ill, or we die       tired or out of breath
              ↓                           ↓
     Exercise builds up our energy  Exercise uses up our energy
```

energy. The fact that they lead (by the rhetoric of my arrangement, not by any logic of the children's own arguments) to a contradiction in meaning on the last line, serves to emphasize the distinctness of the two themes. But the notion of contradiction needs qualification and interpretation when it exists within the alogical life-world stock of knowledge.

This distinction allowed a difference between the boys' and girls' sentences about living creatures to be detected. Where the boys wrote more about themselves and about sport, with meanings which were often on the right-hand side of the plan above, girls did not. They wrote more about energy as health, often considering others or generalities rather than their own sporting experiences. So it seems that the well-documented interest that girls show in other people and in health care has been revealed again in their meanings for energy.

Mechanistic and Political Themes

By teasing out the sense of the non-living sentences a type of meaning emerged related to the uses of energy, to sun and wind, to home heating and electricity, to machinery and fuels for industry. Once again a careful reading of the sentences extracted much more than such a list might indicate. There was, for example, a societal dimension which ran from

> 'Electricity is energy which lights bulbs at home', to
> 'Energy makes electricity which runs the country.'

There was also a tendency to polarise between the renewable and non-renewable forms of energy.

'(Energy) is made from natural things or using natural things, e.g. sun, hydroelectricity, water, wind.'
'Energy is a kind of power or fuel. It can be electricity.'
'Energy is power like electricity, coal, gas and oil.'

There seemed to be considerably more interest in the generation of electricity from alternative forms of energy among the older pupils. In some cases the undercurrent of reference to the public energy debate became more explicit. Eventually it defied all attempts at classification along the original simple lines of animate or inanimate forms of energy. Statements like

'The whole world is short of energy', or
'We are running out of energy'

could only be accommodated by opening another category of meaning — a public issue-based notion of energy. This fourth theme had fewer contributions than the others, and occurred principally among the more able Third Year pupils.

At the time when the pupils wrote these responses, just a few years after the 'Energy Crisis' following the quadrupling of oil prices during the mid-1970s, these sentences stood out clearly as a reference to topical thinking about energy as a shortage commodity. As Meyrowitz had observed, television now routinely introduces children to the full range of adults' concerns, so the sentences served to place the pupils within their contemporary culture. It linked their general knowledge meanings with those of the adults talking in the first chapter, some of whom had shown an instant interpretation of energy in terms of fuel shortage.

Overlaps and Populations

It would be simple enough to count up the pupil adherents for each of the four main themes of meaning that have been identified, but this would almost certainly be misleading. In the first place many of the simple abbreviated sentences were too short and vague to be sure about the intended theme (e.g. 'We need energy'). It is also possible, even probable, that several views are held concurrently by the same individual since they are necessary for exchanges between people. These meaning themes cannot be over-arching so as to be applicable to all situations; in the manner of life-world knowledge each one is linked to a particular set of circumstances.

The sentences, with their different themes and contents, are social constructions just as much as they are personal ones. Most people would have no trouble in understanding each of them and nodding agreement to their use. The over-riding criterion for socially useful knowledge is not a logical structure, but only this wide and agreed usage. The drawing shown

Getting to Know about Energy in School and Society

SURVEY — sentences about ENERGY

LIFE - WORLD
'Provinces of meaning'

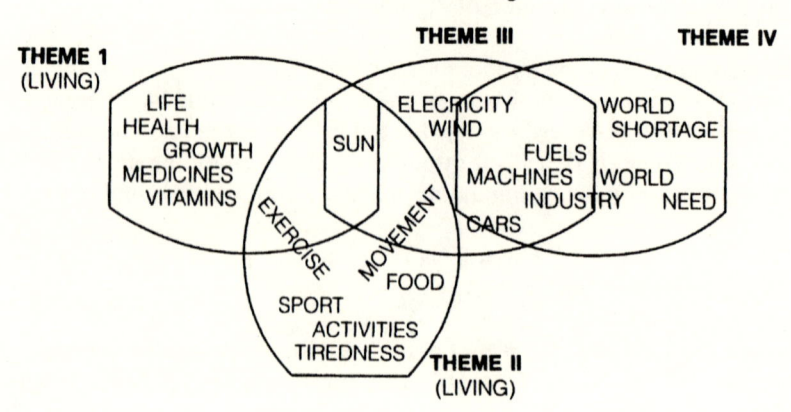

```
                    THEME III           THEME IV
THEME 1
(LIVING)        LIFE         ELECTRICITY    WORLD
               HEALTH    SUN    WIND       SHORTAGE
                GROWTH         FUELS
               MEDICINES      MACHINES    WORLD
                VITAMINS       INDUSTRY      NEED
                      EXERCISE    MOVEMENT  CARS
                              FOOD
                         SPORT
                       ACTIVITIES
                        TIREDNESS
                                    THEME II
                                    (LIVING)
```

above lays out the four themes in a roughly similar fashion to the word cluster generated by the First Year pupils in their efforts to produce a synonym for energy. But this time whole sentences are our data so the meaning themes and their uncertain boundaries can be a little more boldly drawn. The term 'exercise' for instance, spans both theme I and theme II.

'Movement' occupies an equivalent position on the overlap between themes II and III because, as several pupils pointed out, both we and machines can move. And then, as if to show that no two dimensional map can adequately represent all the possible conjunctions of meanings, one Second Year pupil wrote 'Energy lives in electricity', suggesting perhaps an unexpected overlap between themes I and III!

Energy from the Sun

The central position of the sun, on the overlap of at least three themes, is striking. It demonstrated this position through sentences like:

'The sun is the source of all energy.'
'The sun is the source of life and energy.'
'Energy from the sun makes everything live and move.'
'Energy from the sun makes plants grow for our food.'
'We get our energy from plants which get it from the sun.'
'Solar energy can be used to make power.'
'Wind, water and sun are forms of energy.'
'Electricity can be made from solar.'

These pupils had not been taught a science course about energy but they had been exposed to other school influences. On inquiry it turned out that some of these notions may have been transmitted through geography or biology lessons. The geography teacher at the London school had argued that 'All energy comes from the sun' during one of the adult conversations in which, by chance, she took part. (It was disturbing to hear the same teacher talk about nuclear energy in the very next sentence.) In both schools there had been some preliminary teaching about food chains during the Second Year which may help to identify the meaning behind the fourth and fifth sentences above.

We are still left, however, with a very considerable number of sentences, such as the first three, where energy seems to be spontaneously related to an all-powerful sun. Mention of the sun as a source of life is reminiscent of the description of energy as the source of life in the First Year cluster. Perhaps this is similar to the insistence on the pre-eminence of the sun which is shown in the hugely yellow sphere on so many very young children's paintings.

If the term energy has become powerful and ubiquitous enough to be a symbol, it is ready to be drawn into every human interest. Young children of our society may also use it for their own purposes. Sunshine is said, like energy, to be good for you. The young deaf and blind child, Helen Keller, wrote that she had spontaneously equated the sun's warmth with love — an even stronger symbol than energy. Tracking down explanations for meanings in the general culture can go little further than this kind of collection of exemplar material. It is not an exercise in logic, so similarities and verbal juxtapositions are probably as far as we can go without more conversational data. Thus 'the sun is the source of life and energy' tells us more about the felt value of both sun and energy, than it does about some causal link between energy and either photosynthesis or fossil fuels.

Contradictions and the Effect of Context

The term 'exercise' does not just lie on an overlap, it is actually used in what appear to be contradictory ways within these two themes. From the point of view of Theme I exercise becomes an energy/life-enhancing activity. Like medicines and rest, exercise is good for the health of the individual and therefore is said to 'build up' your energy. From the point of view of theme II — human kinetic energy — exercise involves the expenditure of energy which was stored in the body through the food that was eaten. Just as a car uses energy from petrol in order to move, so we 'use up' our energy when we take exercise. Both these phrases 'exercise builds up energy' and 'exercise uses up energy' made frequent appearances in the children's writings. For anyone trained in the natural sciences they had to seem downright contradictory.

Implied contradictions have stimulated crucial experiments in the history of science. If conflicting predictions were made it seemed that one experiment could, at a blow, decide between the two theoretical positions. The eighteenth century dispute about the nature of light made contradictory predictions about its speed in glass. The twentieth century confrontation between Newtonian and relativistic theories of gravity was put to the test once it was clear that stellar observations during an eclipse would decide between them. With hindsight the experimental verdicts may not seem quite so impartial and decisive as might be. The wave theory of light was not to survive in its simplest form for more than 60 years despite the favourable verdict that the speed of light was reduced by glass. The deviation of star light by gravity which supported relativity was little more than one half of what Einstein had predicted. However, the logic of crucial deciding experiments was accepted on both occasions.

In the present research the crucial connection between 'exercise' and energy is less decisive because socially constructed knowledge tolerates logical contradiction. Several of the classes were asked, either orally or on paper, which one of the statements — A or B — they agreed with.

A   'Exercise builds up your energy.'
B   'Exercise uses up your energy.'

One small First Year girl abolished the whole notion of decisive contradiction right at the outset. Asked whether she believed A or B she answered in class, 'It could be both', and her friends all agreed. At the other end of the range two top ability Third Year classes, also asked to decide between the statements, gave a majority vote for the compromise: 'It could be either, it just depends on the circumstances.'

It seems that a child, or even an adult, may hold such apparently contradictory views because they apply them in different contexts. The fullness of the occasion in terms of intentions, physical context, and personal involvement will evoke roughly similar meanings from all participants in any culture. Contradictions simply will not surface while the situation 'cues' the appropriate meaning theme. Given another situation, concerned perhaps with someone taking exercise to recover from an illness rather than wearing themselves out running round the playground, the other theme of meaning will be simply appropriate and not contradictory at all.

> ...elements of knowledge which are theoretically self-contradictory need not necessarily conflict in the natural attitude. The origin of the theoretical contradiction derives from the heterogeneity of the situations in which the knowledge was acquired. (Schutz and Luckmann, 1973)

When a child is isolated from others by the demands of a questionaire the natural consensus which drives sociable talk and defines the meaning context, is missing. It might be expected that some kind of logical inertia would now ensure consistency of meaning. However, that was not always the case in this study. A moment's thought between each sentence, or even the odd word mid-way through one, seemed to be enough to call to mind another situation along with its own contextual meanings. Here is one such set of consecutive sentences from a Second Year pupil:

> 'Energy is when you build up power and strength and this helps the muscles move more freely.'
> 'If a person is sick and can't get up she/he hasn't got enough energy to move.'
> 'So when we do exercise we use up lots of energy.'

It looks as if the first meaning for energy to come to mind was that of Theme I. In the second sentence the beginning suggests that this meaning of health and life was still dominant. Maybe it was the final word 'move', which triggered thought into Theme II, that of human kinetic energy, and caused this pupil to construct a third sentence whose meaning might seem to be in opposition to that in the first sentence.

The perspective of 'common sense' simply accepts the whole range of meaning variation. Individuals may even be able to reflect colloquially upon this variety without finding the contradiction problematic. In the following extract from one of the adults' conversations, another aspect of the alogical rhetoric of everyday life — the common proverb — is used to smooth out any apparent contradiction. She was asked whether exercise built up your energy or used it up.

> 'I suppose if you have more exercise I should imagine you would be more energetic.... Swings and roundabouts, probably, the more energetic you are the more exercise you would get.'

Quality, Quantity and Concept

The word 'energy', as we have seen, slips easily into the meaning of 'energetic'. The new term is an adjective instead of a noun, and it has a far stronger human connection than has 'energy'. Even if we substitute 'energeticness' in order to make another noun for comparison with 'energy' the human connections are unchanged; even something of its adjectival flavour still remains.

The philosopher John Locke used the term *quality* to signify words like 'energeticness' and 'redness' which are formed in this simple way from the corresponding adjectives. Although some kind of generalizing process must have taken place for such abstract notions to have been formed, they are not yet true concepts. In several important ways these nouns are little more than the adjectives from which they were formed. A scientific concept like energy will possess new characteristics of its own. The two most important of these are measurability, and some independence from human sensibility.

There are a number of other quality/concept pairs which show these characteristics, for example, size and volume, heaviness and weight. The concept member of the couple, the second one, is invariably sharper in its meaning. That allows the term to take its place in some simple piece of theory and so be measured in a comprehensible and replicable way. Weight and volume are carefully defined in science and are used quantitatively. They are involved in calculations relating respectively to the equilibrium of a lever, or to the length, breadth, and height of an object. Heaviness and bigness are qualities which do not figure in any calculations; there is no need for them to be quantitative. They can certainly be placed in rank order by human assessment; we may say that one thing is bigger or heavier than another, but this is not measurement. The lack of measurability of a quality is directly connected to its rough and ready detection by the human senses.

This insight can be applied to the pupils' understanding of energy, and its possible separation from energeticness. For this there is no need to rely entirely upon the evidence of freely chosen sentences; a direct question could be asked about the quantification of energy. Immediately after writing their sentences the Third Year pupils had to answer whether they thought that energy was 'the sort of thing you could measure'. (In neither school had this been taught in any way.) Several of the pupils wrote more than a mere 'yes' or 'no'. Some suggested measuring it by how much weight could be moved, or how fast a person could run. The real interest of these results emerged when they were compared with the human or non-human meanings of energy in the same children's sentences.

The results showed that only 28 per cent of those who gave entirely human associations for energy thought it might be measurable: whereas 72 per cent of those giving at least one non-human example of energy thought it was the kind of thing that could be measured, even though they had not yet been taught how to do so. It is tempting to infer from this that children who had an almost exclusively human view of energy were thinking about the quality of energeticness rather than energy. It may have been more of a personal feeling of well-being than the sort of p-prim on which a concept could be built.

First Stage in Generalization

More insight into the nature of concept forming could be gleaned from the answers to the first question 'What is energy?'. A concept should embrace more than one manifestation of energy, either by listing or generalization. The simplest generalization requires two simultaneous processes: there must be a mental comparison of different energy situations, and this comparison must yield a common feature.

Children's Ideas on Energy

In the most primitive case of just beginning to generalize, two different examples of energy were simply listed side by side. Where there was no effort to extract and name the common feature, as in the sentence below, true generalization had not yet been achieved.

> 'There are two kinds of energy. One is the kind we get from food, and the other is electricity'.

The next example shows a slightly improved version, which may just have crossed the boundary into effective generalization.

> '(Energy is) a source of power which makes things grow or makes electricity.'

Here the common feature is the vague term 'power' which is little more than a synonym for energy in general parlance. What is required in order to decontextualize the notion thoroughly, and to link together its appearance in different circumstances, is some general idea of *how it operates*. There were signs that this process of naming what energy could do, took place within a single context before an attempt was made to generalize across different situations.

> 'Energy is electric current, that is power. This power helps things *to work*. Such as a light bulb.'
> 'Energy is something we *use to move* things.'

These two sentences use the words 'work' and 'use' to describe the salient feature of energy. Still they stick to a single context, to electricity or human effort. In really successful generalizations it was the same two words which proved to be the most successful operational terms with which to make complete generalizations.

> '(Energy is) A kind of force that helps things move.'
> 'Energy is power; a thing that moves on its own uses energy.'
> 'Energy is a thing that everything needs to work.'
> 'Energy makes things work.'

Three out of the four of these efforts began with a predicated noun — 'force', 'power', a 'thing' — which is a synonym contributing almost nothing to the generalization. It was the *verbs* — *move*, or *work* — which provided the important common feature.

Getting Better at Concept Formation

Research has found similar results in fields unrelated to energy. Students' definitions of force (Leboutet-Barrell, 1976) of a chemical acid (Kempa and

Table 2 Percentage of Pupils Making Successful Generalizations

First Year	8%	
Second Year	25%	(Top group 42% : Lower 4%)
Third Year	33%	(Top group 51% : Lower 11%)

Hodgson, 1976), and of electricity (Solomon et al., 1987) all showed this same developing emphasis on the use of verbs to define scientific ideas. In the first of these a dramatic increase in French childrens' use of verbs took place during the age-range corresponding to our First to Third Years. Such a facility with verbs would then be one factor which would lead to a steady increase in the number of pupils who could make generalizations about energy. Other effects which would also encourage generalization might be a growing familiarity with other over-arching concepts either within school or outside, and a greater capacity to express abstract ideas in words.

Since the Second and Third Year groups in both schools had been placed in ability sets for science it was also possible to see that this emerging ability to generalize occurred far more often in the top ability group, than in the other (Table 2). Another way to show this close connection between the use of operational words in generalizations as a first step towards concept formation, is to set a task which calls for words to describe 'what energy does'. A class of Second Year pupils in a rural English school who were just beginning their first course on energy, were asked to form small friendship groups. This enabled some pooling and reinforcing of social knowledge. Each group was then given a double sheet of paper to fill in with their ideas about energy. On one side the heading read *Things which have energy*, on the other was the heading *What energy can do*. The children talked together and filled in their ideas on the two sides of the paper. After ten minutes the sheets were collected.

The results were very like those obtained from the free sentences. Groups of less able pupils filled in a long list of *Things which have energy* — fires, batteries, lights, torches, people, engines, etc. — but they seemed unable to make much progress with *What energy can do*. The groups of more able pupils had little difficulty with what energy does and wrote sentences like 'Energy makes things move', or, 'Energy powers things.' Once again the verbs dominated.

There was an interesting extra feature in this exercise. The two most able groups who had filled in sentences about what energy can do, wrote almost nothing on the *Things which have energy* side of the page. They did not seem to have left this out because they were too difficult, nor because of lack of time, since they only wrote three or four 'doing' sentences and then passed the time in idle talk on something else. When they were asked why they had left the other side of the page empty, pupils in both groups replied

'It is all there in those four sentences on this side.' They pointed to what they had written about what energy does, recognizing in the generalizing power of the statements made about the operation of energy as an economy of abstraction.

Postscript through Children's Drawings

The work of some psychologists has suggested that there may be 'visualizers' as well as 'verbalizers' in the population but their research has always focused on the solving of verbal and pictorial puzzles rather than asking for the formulation of meaning. The written answer to a question like 'What is energy?' must favour the verbalizers for it is hard to see how children could be expected to draw the energy concept.

But children could draw some of the effects of sunlight (p. 58), and this would make a valuable comparison with the work on energy in two ways. The central position of the sun in children's ideas on energy makes this of contextual interest. More interestingly, light is a concept which could be written about in words, as 'energy' was, as well as forming the topic for drawing. This might show if asking for written answers did favour the verbalizers at the expense of the visualizers.

Some Third Year pupils in the rural English school, who had not yet studied light, were asked 'What is light?', as in the main energy research. Then they were asked to draw a picture of a sunset *showing as much as they could about light.*

Some of the children gave particular examples: 'Light is daylight', or 'Light is a bright thing', and a few could not answer this question at all which is, perhaps, not very surprising. A very few produced operational statements such as 'light is what makes things bright for us to see.' Their drawings also showed either bright objects (mostly the sun) or, in what seemed to be an operational vein, indicated the effects of light correctly observed through shadows or one-sided illumination of objects in the landscape.

There was a significant numerical association between the operational nature of the verbal response, and the same operation of light in the picture of $+0.79$ $(+/- 0.11)$. This is enough to show that in this sample of 80 children most but not all of the visualizers were also effective verbalizers. That gives some confidence that asking for written statements about energy has not penalized the children with a more pictorial style of exposition.

However, the ability of the children, as measured by their grade in the yearly science test, showed much less correspondence with their success in the pictorial task. One way to understand this lack of association is to see this, not as a test of the learning of science, but rather of untutored ideas about what they had observed.

Getting to Know about Energy in School and Society

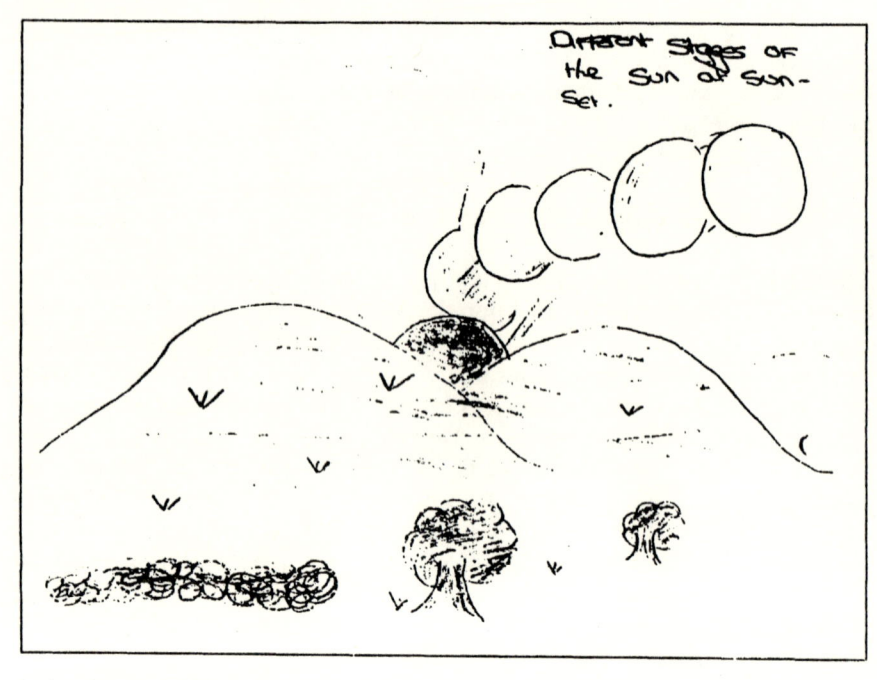

1   Landscape without light

2   Landscape with light

Chapter 4

Evidence from Talking

Talk as a Research Tool

Talking produces quite different text from writing. This chapter will probe carefully into classroom talk of different types to see if more information on children's ideas about energy can be extracted from them.

Almost all the answers used in the last chapter were given piecemeal by lone children who often crouched over their answer sheets to protect them from prying eyes. Other children did the opposite, and kept trying to steal a glance at what their neighbours had written. The common school situation of writing responses had reminded them of tests and examinations in which it was important to get the right answer, and to do as well or better than the others.

Talking will be differently affected from writing, but situations cannot escape carrying messages about meanings. If the teacher begins the talk it may have all the overtones of trying to teach, and trying to learn, the 'correct' answer. Edwards and Mercer (1987) described primary children's classroom talk with their teachers as:

> ...a rather one-sided affair in which the teacher's role as an authoritative bearer of the ready-made knowledge simply finds alternative, more subtle means of realizing itself than the crudities of brutal transmission. (163)

Such transmission, brutal or not, of accepted knowledge would throw very little light on the pupils' own ideas. Teachers often call the question and answer sessions by which they lay down the dominant ideas for a new piece of learning, a 'discussion session'. Generally it is certainly not discussion in the accepted meaning of the word, where there is an interplay of meanings and ideas. To obtain research data from discussion in the classroom quite a different context will need to be developed, and the word *context* itself needs to be expanded in several dimensions if it is to take into account the multitude of cues which affect both meaning and behaviour.

Context and Meaning

The classic psychology experiment to establish the effect of environmental context involved having matched groups of subjects learn random lists of words on dry land, and deep under water, with the subjects kitted-out in cumbersome diving suits. It was then soberly established that recall was significantly better when the subjects were in the same environments as those in which they had first learned the lists.

The school classroom is a less exotic environment, but it conveys many messages. These are not the same for all pupils in the room. The previous histories of pupils within the classroom, their successes and failures there, will greatly alter the range of possible attitudes and actions. That fact alone suggests that context is also a private and personal affair inside the head, and not just one of physical environment.

Classrooms are also the site of social groupings and social acts which obey special function-related rules. I learnt this through classroom teaching, as all teachers do, but most acutely when carrying out a rather uncomfortable piece of research on classroom behaviour (Solomon, 1988). During one whole school year I engaged in a serious but solitary pursuit of the rules of meaning which might make sense of behaviour in secondary school science laboratories. Being myself a part of the social scene I was thrown into two distinct roles, and hence two different contexts with different meanings. During the lessons, as teacher, all I noticed would have to be recorded, and this itself heightened awareness of the pupils' reactions. When the bell sounded I rushed off, now as researcher, to write down notes on what they had said and done at different moments in the lesson. Myself as teacher was also watching the researcher's notes and appraising them in terms of the meaning of a 'good lesson', or the reverse, which could be read into them. It was a curious and uncomfortable process. Erwin Goffman's performance theory (1959) might have described it as:

> ...not unlike a man desperately trying to play tennis with himself. Again we are forced to see that the individual is not the natural unit for consideration but rather the team and its members.(149)

The data had to encompass the whole 'classroom team' as context. Teacherly feelings about how the lessons were going were not to be ignored, they were a part of the behaviour 'team'. Adolescence is a time for special group rules and roles, so secondary classrooms fit particularly well into the performance approach to context. Aspects of this will become very important in both setting up and analyzing pupil talk.

The notion of *context* which so powerfully affects the meanings we attach to words has thus expanded from that of the physical environment, through the consideration of personal histories carried around in pupil's heads, to the teacher's professional role and the social rules of group

Evidence from Talking

performances. The context which triggers meanings is also dependent upon people's perceptions of each other's reliablility and intentionality in their performance. We have to be sure that others have read the context in the same way as we do.

Roles, Performances and Cheating

Social pretence, especially when it is detectable, forms no basis for meaningful talk. The contract of classroom performance does not easily permit the teacher to change role surreptitiously. The accepted behaviour of teachers is to ask questions; the accepted nature of such questions is that they have access to the 'right answers'; the accepted rights of pupils include being told when their answers are correct. As the unfortunate extract below shows, such conventions may not be flouted with impunity.

> The teacher starts the discussion by asking 'What is energy?'. After some minutes of questions and answers, during which the teacher tries to make only non-committal responses in the approved manner of educational research, the following occurs:
>
> P1. ...When resting you are getting your energy back.
> T.   *That's interesting...*
> P2. Stop commenting on us.
> P3. Eating gives you energy, isn't it?
> T.   *Eating? And resting also?*
> P4. Yes, of course it does.
>   Chorus of several Ps. Yes, Yes...
> P2. Do you know when we're wrong, Miss?
> P3. (challenge straight to teacher)...WELL, WHAT IS IT?

If genuine inter-pupil discussion is to take place new kinds of classroom performance will need to be negotiated, slowly and with care. The pupils will feel too uneasy to make meaningful statements if they do not understand the teacher's behaviour. The teacher's intentionality, as well as the rules and roles to be adopted, are an integral part of the context; should they change then the whole meaning environment will change too. Sentences about the meaning of energy which would have carried social and personal approbation in the pupil-teacher learning mode, could be made worthless, the subject of joke or subterfuge. A new performance can only happen when pupils have learnt its rules, and can feel what actions and meanings are appropriate.

Once established, such talk should be able to build usefully upon the themes of meaning already mapped out. If it does map on to these that will provide some validation of the method, as well as an extension of our understanding. Many of the written sentences reported in the last chapter were terse and ambiguous. Because of the curious nature of common-sense knowing (chapter 1), cross-examining each child to find out the exact meaning intended is not the straightforward process that some researchers have assumed. Meanings are essentially shared so, in a suitable context, pupil talk should be able to expose more of the structures of particular meanings than private writing.

Kinds of Talking

Jerome Bruner, that provocative educator and researcher, believed he had evidence that children used language as 'an instrument of thinking' in different ways at different ages. He charted the path from its early use for pointing or reference, to the final stage of constructing symbolic representations. He was also an enthusiast for the use of oral and written langauge in the teaching process (1978). He wrote that it is:

> ...not only the medium for exchange but the instrument that the learner can then use for himself in bringing order into the environment. (6)

Of the several useful kinds of talk one is, paradoxically, both silent and private. This is used for internal problem solving. The Russian psychologist L.S. Vygotsky, watched small children trying to solve the kinds of simple practical problems that Kohler had previously set for his apes. They had to manipulate sticks and boxes in order to reach some food. What struck Vygotsky most was the 'planful' nature of the continuous chatter which accompanied the actions of 4–5-year-old children working on their own.

> Children solve practical tasks with the help of their speech, as well as their hands and their eyes. This unity of perception, speech and action, which ultimately produces internalization of the visual field, constitutes the central subject matter for any analysis of the origin of uniquely human behaviour. (Vygotsky, 1962)

Older children and adults do not care to be caught out talking to themselves while working out their ideas, but sometimes they use psuedo-discussion for just this purpose. When a group of pupils is encouraged to use talk to explain events this private problem-solving commentary has a chance to resurface. We might expect it to be incomplete, tentative, and almost non-communicative in intention.

At the other extreme much meaning-related talk has, as its main purpose, getting confirmation of what we already know. The feedback is all-important. The talk may be a re-living of events from the immediate past. In between lessons and at play-time, school children congregate in small groups or pairs to 'discuss' a television programme which they have all heard, or a 'thing' that happened to one of them. The compulsion to talk might be no more than a vanity, but often this social rehearsal of the familiar but recent appears to fulfil a semantic need. It is as though every experience has to be talked out in assenting company for its meaning to be assimilated by the speaker/actor. 'Isn't it?', they say continually.

The essence of such talk is getting others to agree. But all conversation is a two-sided performance for which the rules require alternating responses. The side which has been listening is under some pressure to take their turn at speaking. One kind of such sociable response is adding to what has just been said. This serves at least three important functions. It reinforces the first speaker's ideas (possibly by expanding them), it keeps the argument going (often by additional examples), and it enables the respondent to be a part of the social performance.

Some responses are merely designed to keep the talk going. This is a mutual 'patting' affair (dubbed 'phatic communion' by the sociologists) in which there is no communication of new ideas. Words may even give way to grunts or nods which make it possible for the first speaker to go on. It was the anthropologist Malinowsky who first drew attention to its rather odd intention.

> ...much of what passes for communication is rather the equivalent of a handclasp or an embrace; its purpose is sociability;...it is a way of keeping the channels open even when no information travels through. (Bolinger, 1975)

Refuting others' ideas is a rarer and stronger kind of response. Most people do not readily speak out against the majority in an on-going argument, although the work of Moscovici (1976) cautions against assuming that final agreement is produced either exclusively by the majority, or by an over-bearing member of the group. Minority influence can be strong, but both it and majority influence may lose their grip once the group has dissolved. In the conversations taking place within adolescent peer groups the wish to appear to agree is especially powerful (Marsh et al., 1978). Disagreements point either to latent personal hostility or to an unusually well thought out and committed point of view on the part of either a group leader, or a group outcast who has nothing to lose.

Getting Useful Talk Going

Some of the first research to use children's work-related talk was carried out by Douglas Barnes in 1976 and in collaboration with Francis Todd in 1977. He used four very small groups of children who were given tasks to carry out by discussion. There were between two and four children in each group, and no teacher was physically present although the work to be done had been set, by the teacher, to build upon recent class work.

The transcripts make fascinating but rather tantalizing reading. Such examples of learning through talking (as Vygotsky and Bruner had claimed that children do) go a long way towards validating Barnes' contention that the school curriculum could be reformulated so as to allow children to organize their own learning by discussing in groups. Other excerpts from Barnes' transcripts are less convincing: almost no progress in explanation is made. Often the problem is that the discussions terminate too soon. In his perceptive comments on the children's talk, Barnes attributes this 'closed approach' to either premature consensus, or to ritualized question-and-answer put on for the benefit of the unfamiliar tape recorder. This ritualized approach, where one pupil becomes a quasi-teacher by both reading out the question and deciding upon the 'correctness' of the answers given, may be encouraging and confidence-building, by making the unfamiliar leaderless situation more like that of the normal classroom. However, it too is a social pretence, and not true pupil to pupil discussion.

The presence of the class teacher may go some way towards solving this particular problem since no child, however bossy, could then usurp her position. But the hard question here, in view of the disastrous effects of breaking the accepted norms of intentionality already described, is whether it is feasible for the children to hold their own discussions with the teacher present. It is certainly not a usual event in the classroom.

The method adopted for this research was to station a tape-recorder, in full view of the children, on the teacher's bench for every lesson. Its presence and purpose were explained and gave rise to considerable hilarity at first, with frequent demands to replay the recordings. Only after a full school term of visibility, during which the recorder had become an unconsidered part of the context and performance, was it used for the serious recording of teacher and pupil talk. During this time the teacher also instituted a regular ten minute discussion period at the beginning of every lesson. Gradually the context became familiar and the performance was learnt by all the 'teams'.

There are some statistical methods of analyzing discourse which can be used to show whether or not discussion is truly pupil-pupil centered. The use of a modified Flanders' Interactional Analysis (1970) has been described elsewhere in some detail (Solomon, 1985). The plans opposite show the typical charts obtained when the teacher conducts a one-sided question and answer 'discussion' (Chart 1), and the quite different pattern which describes the concentration of pupil-pupil interaction in a more valuable kind of discussion (Chart 2). For purposes of validation, this kind of analysis provides 'hard' proof that it is not just the kind of teacher-led talk which Edwards and Mercer reported (quoted page 63). From the interactionist point of view, these charts also suggest how meanings will be constructed. In

Evidence from Talking

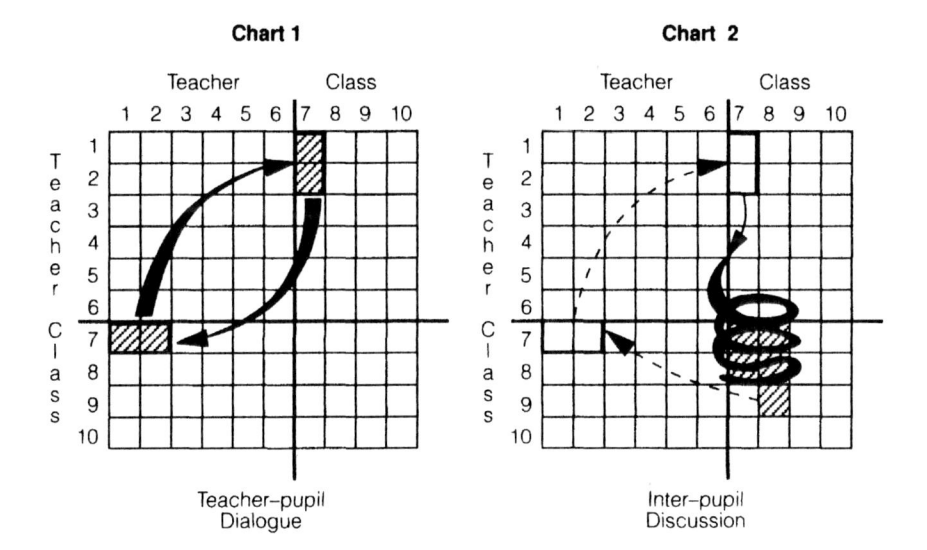

Chart 1
Teacher–pupil Dialogue

Chart 2
Inter–pupil Discussion

a chart 1 type of discussion the pupils will be trying to construe the teacher's meaning as they search for answers which will satisfy her questioning. Only in the second kind of discussion will the pupils fall into the more familiar kind of behaviour where they make contributions which use meanings they feel sure their friends will recognize. They are now talking valuably from within the mini-culture which generates their common stock of general knowledge about energy, and other matters.

For further validation of classroom talk as a research tool, we can also use recognition of the familiar traits of unconstrained talk. The four categories of talk identified in the last section may be expected to surface. Children's ideas will be offered for general agreement: most pupils will go along with them either expanding or confirming them. Occasionally we should hear refutations. There may even be occasions where the slow working out of an idea suggests the operation of Vygotskyan problem solving through reflective 'inner speech' made audible.

Talking for Consensus

The first two extracts below come from different Fourth Year classes and have been chosen to show the general characteristics of cooperative brainstorming discussion. In both there is a potential element of disagreement (indicated*) and yet, through the comfortable conventions of social talk, almost all of these extracts contrive to finish on a consensual note.

   ...*Teacher.* What is energy?
   P1 and 2.   Power.

65

Getting to Know about Energy in School and Society

Ps.	Power, power, driving force.
P3.	What makes something work?
P4.	Something that originates from the sun.
P5.	Power that enables us to work. You need energy to work.
P6.	It comes from the sun.
*P1.	Not everything comes from the sun — like petroleum.
*P7.	And nuclear.
P5.	Power is....
*P8.	That has come from the sun too.
P9.	And our energy.
P2.	And hydroelectric power.
	(...interlude as teacher quietens two noisy pupils at the back...)
T. (to P9)	What were you saying?
P9.	We've got energy as well.
Lots	...Yes, yes, food, food.
*P8.	That comes from the sun as well. It does.
P5.	Chemical energy.
P10.	Food.
*P8.	...plants indirectly from the sun but you need energy to make it grow.
P3.	Sun makes the plants grow and...we eat...
P2.	Or the animals are fed from the plants so we get...
P3.	And then we eat them.

This typical classroom discussion is predominately agreeable, as is shown by the 'chaining' effect of the comments. The majority of the pupils simply pick up a previous comment and add another modicum of knowledge to it. Quite a few of the contributions even start with the word 'and'.

Nevertheless, there is a disagreement hidden in the extract. Pupil 6 is the first to make the point about all energy coming from the sun which the argumentative pupil 8 sticks to, during the rest of the extract. Pupils 1 and 7 disagree with this and yet no protracted adversarial argument ensues. It is not so much that pupil 8 wins the day but that the sweep of sociable talk simply ignores the incipient controversy.

In the second extract the teacher is heard intervening in order to sharpen the controversy by forcing the class to vote on the issue. Even this potentially divisive strategy does not prevent the pupils from arriving at an agreeable consensus.

...Teacher.	*How do human beings get their energy?*
P1.	From food.
P2.	Water.
P3.	Oxygen.

66

P4.	Nitrogen.
P5.	From exercise.
*P1.	By doing exercise.
*P6.	You lose it by doing exercise.
Teacher.	*Let's have a show of hands. How many people think you gain energy by doing exercise...TWO. And how many people think you lose energy by doing exercise...SEVEN.*
P5.	You may lose energy.
P7.	It makes your body fit.
P5.	Yes, I think *both*.
Teacher.	*How many people think it could be both?...FIVE.*
*P6.	It improves your muscles but it doesn't actually give you energy.
P7.	It makes your body better able to support you.
P8.	Energy is in our muscles.
Lots	Yes, yes.
P3.	Muscles store the energy.
P2.	Exercise makes your heart stronger.

Here the original confrontation is between pupils 1 and 6. The numbers recorded by the teacher in her 'hands up' vote is far less informative than the next part of the discussion. Pupil 5, who first mentioned exercise, is heard moving over to a position where both points of view are acceptable. Then, despite the teacher's tally and the careful point of distinction being made by pupil 6, the context of fitness/health draws the whole of the group into apparent consensus. They all recognize both meanings and decline to be drawn into any exclusive definition.

Most of the next short piece of Second Year discussion is anything but sociable. It took place between two pupils who did not like each other and let this feeling spill over into refutations about energy and exercise. This episode is particularly interesting since the writing of P3 in the paper tests (reported in the last chapter) showed that both the health/fitness meaning and the human kinetic energy meaning were known and used. This time, in talk with the class, the animosity of argument brings out a straight denial of this point of view.

...Teacher.	*What does energy mean?*
P1.	Strength.
P2.	Something that helps you do something like running.
*P3.	When you run you build up lots of energy.
*P4. (Sharply)	No, you need lots of energy to run.
P5.	You need energy to live.
*P3. (Emphatically)	It builds up every time you do something.

67

Getting to Know about Energy in School and Society

Teacher.	'Energy builds up when you do something' is that right?
P5.	Yes, you need energy to live.
*P4.	No, no.... (drowned out by clamour)

Despite the unresolved nature of this dispute it is interesting to hear P5 intervening twice in a way that seems to be intended to restore the sociable conventions of normal conversation. The sentence that he uses on two occasions, to make the peace, is precisely the one to which the adherents of these meaning themes could both agree. However, the argument is far more personal than thematic so this olive-branch does not reconcile the angry opponents. Conversation, even when it is about the meaning of a physics word like 'energy', can display the whole human range of intentions and emotions.

The last extract comes from another Fourth Year discussion about energy. It all began with one pupil defining energy as 'a source of force or a source of power', to which another pupil countered, 'Why does it have to be a source?' This was a common topic in these discussions as we shall see.

The verbal challenge produced no immediate response but, a little later, one of the girls burst out with an uncharacteristically long contribution. The talk had moved on to the subject of food but this pupil seems still to have been worrying about the idea of a 'source of energy. Now it became important to gain corroboration from the others for a personal point of view. Unlike the majority of contributions this one came out with almost a tone of urgency. An important problem was being solved with a rather jerky piece of language, and offered to the class for recognition and confirmation.

'...Like uranium, it hasn't got any energy in it at all. You have to do something to it to get energy.... It's nuclear like energy.... I mean, it sounds a lot of energy when you do something to it; but when it's in a natural state it's just rock or something. It don't do anything.'

These extracts show many of the features of talk which had been outlined and were expected. They also show the expression of a range of pupil feelings which might not have been expected in response to such a dry question as 'What is energy?'. Social interactions, however, are bound to be full of social feeling because that is part of the context of talk — far more so than it is for writing. The extracts have been quoted, in part, to demonstrate this very point. We recognize unstilted children's talk by just such criteria; had these features been missing the talk would have been unconvincing to our skilled but commonplace understanding of social exchanges. Most important of all, these affective overtones must be welcomed in any search for social meanings, since they will have been a part of the context in which the social stock of knowledge was laid down and is continually reinforced,

Evidence from Talking

whether this is home (as in the Nigerian pupils' mother-tongue meanings for energy, page 35) or in the social arena of the classroom.

Trails and Topics

The 'chaining' effect of the children's contributions to discussion suggests that whole phases of it might be laid out in trails of slightly abbreviated comments. This would show the semantic order in which clues to social meanings arose, and then were followed up.

When this process was carried out on the classroom transcripts the trails were sometimes completely linear, with each comment following on the heels of the one before. But the time-scale of reflection or inner speech does not always conform to the real time of discussion so that, in most trails, it was also possible to see where a reflective pupil might be responding to an idea given earlier. Indeed, the cumulative effect of several of the topic trails is a mass reaction of those who talk and those who just react, ending in a clamour of unresolvable contributions.

Extracts given so far show two of the notions that were also found in the written answers. The energy/exercise contradiction arose in two of the extracts, and the sun as provider of all energy turned up in another. The following trails all come from Fourth Year discussions and, as might be expected from our Third Year evidence, the urge to construct a generalized operational statement about energy turned up in five out of six of the classes questioned. The following example is a specially simple trail where this objective — to abstract and universalize the energy concept — remained strong most of the way through. (Radiating lines indicate that several voices were saying the same word simultaneously.)

'What do you think energy is?'

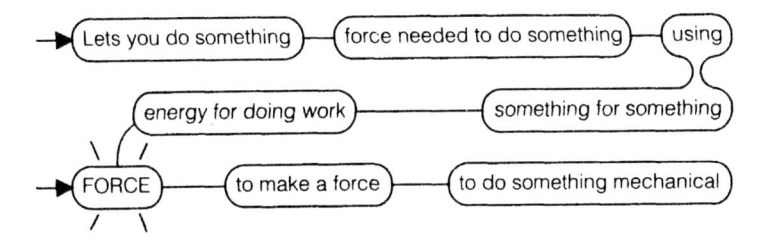

```
→( Lets you do something )—( force needed to do something )—( using )
      ( energy for doing work )————( something for something )
    \ | /
→( FORCE )—( to make a force )—( to do something mechanical )
    / | \
```

All the orthodox physics definitions of energy refer to the concept of 'work', and sometimes also to 'need' in connection with work, so it is reassuring to find that the socially constructed meaning for energy has room for something similar. None of these classes had yet studied the work concept in physics, although all had just completed a unit about force.

Getting to Know about Energy in School and Society

Other findings, which might have been expected from an extrapolation of our analysis of written responses, were that non-living examples were now more common than they had been in the written answers of younger children, and that the attempt to generalize was found more often in the company of the mechanical or non-living than alongside purely human examples. (In the first extract of the last section the emphatic statement from pupil 9 — 'We've got energy as well' diverts the thrust of argument away from abstract generalization and towards food and food chains.) Both these findings confirm and extend the impression, already given in the previous chapter, that development away from the egocentric or personal seems to precede the ability to generalize.

Can Energy be Stored?

There was valuable evidence from these discussion trails about energy 'source' and 'store', which had made an appearance in a few of the written answers. Several pupils had answered the question 'What is energy?' by writing phrases such as 'a source of power', but the significance of this was hard to gauge. The Third Year pupils had been asked the supplementary question, 'Do you think energy can be stored?', but their written answers had been too terse to give much insight and did not seem to correlate with any other features (*see* chapter 7). However, in these discussions the topic of storage was very popular and gradually the sense of what the pupils had in mind emerged with a clarity and lack of prompting that few other methods of research could have equalled.

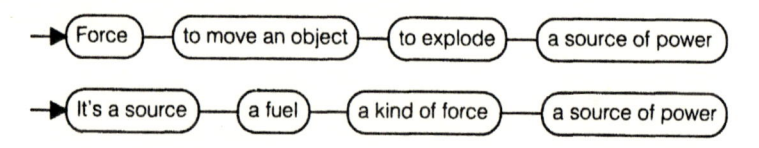

→ (Force)—(to move an object)—(to explode)—(a source of power)

→ (It's a source)—(a fuel)—(a kind of force)—(a source of power)

These two trails came from different classes and both demonstrate that the consensual power of the final phrase closes down all further discussion. Fortunately there were other trails which produced more illumination.

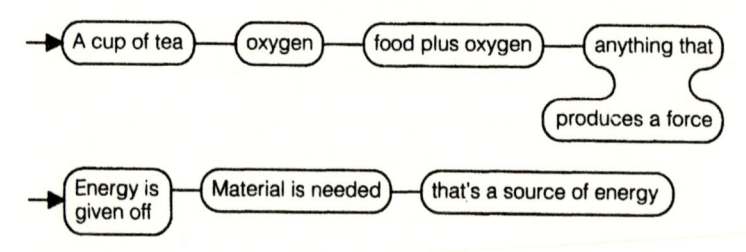

→ (A cup of tea)—(oxygen)—(food plus oxygen)—(anything that produces a force)

→ (Energy is given off)—(Material is needed)—(that's a source of energy)

70

Evidence from Talking

The next trail in this discussion was triggered by the teacher's question 'What sort of things have energy?' With the idea of a 'source of energy' still fresh in their minds the pupils used the phrase again but now in a slightly different context. This begins to make clear why the comment 'a material is needed' was made.

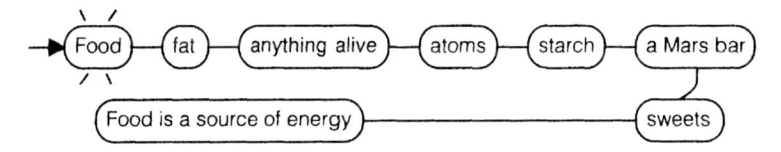

→ (Food) — (fat) — (anything alive) — (atoms) — (starch) — (a Mars bar)
(Food is a source of energy) — (sweets)

It seems that 'a source of energy' can mean a material — food or fuel — to some of the pupils, although the phrase originally appeared after the abstract generalization 'anything that produces a force'. Can 'source' have a meaning which is relevant to the abstract nature of energy? In the following two consecutive trails the phrase is used, challenged, and answered, without any help from the teacher.

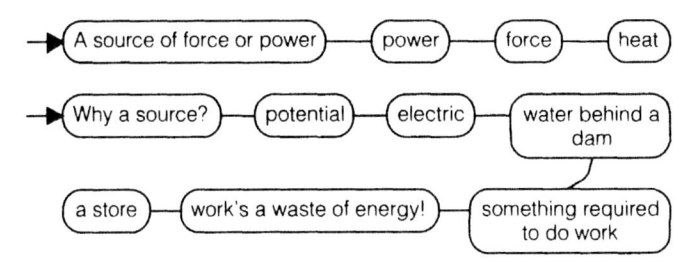

→ (A source of force or power) — (power) — (force) — (heat)
→ (Why a source?) — (potential) — (electric) — (water behind a dam)
(a store) — (work's a waste of energy!) — (something required to do work)

This able group of pupils have managed to achieve what Barnes calls the 'open approach' where they make progress by questioning each other. They also demonstrate that the term 'source' can be used in an abstract generalizing sense ('something required....') as well as in a concrete illustration such as a dam. They even make a rather good pun.

The question for the listener is why energy, if it is a power, or what is needed to do work, is so often qualified by this term 'source'. The answer begins to make an appearance in a rather noisy trail from another class who were discussing how a person gets energy. Food, drink, sleep and respiration had all been proposed in answer to the question, with the majority voting for the first and last of these.

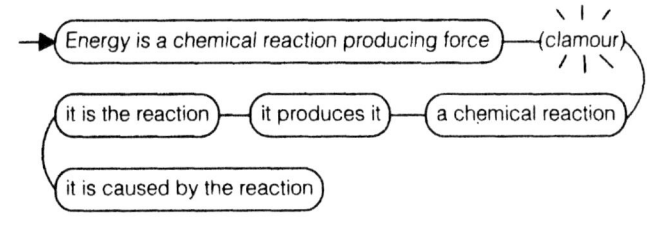

→ (Energy is a chemical reaction producing force) — (clamour)
(it is the reaction) — (it produces it) — (a chemical reaction)
(it is caused by the reaction)

71

Getting to Know about Energy in School and Society

This class were unanimous that energy could be stored in food and also in petrol. There was no problem there. The disagreement was between the 'explosion' line of thought which held that the chemical reaction was itself the energy, and the vaguer idea that the reaction simply produced or caused energy. A related problem was the sense in which you could say that petrol or food actually stored energy if, as several pupils commented, 'It *only causes* energy'.

A class vote showed just how widespread this confusion was. By nineteen votes to three this class of more than average ability Fourth Year pupils decided that 'energy is caused by food and petrol' was more true than 'Energy is stored up in food and petrol'.

The point is particularly well made in the final part of a long trail from yet another Fourth Year class. Once again they are discussing whether and how you can store up energy.

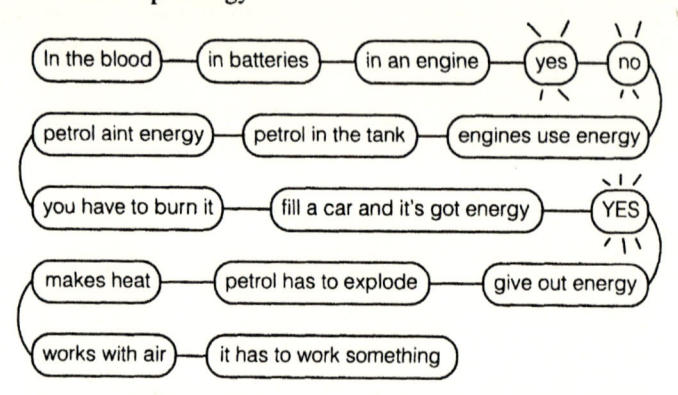

```
(In the blood)—(in batteries)—(in an engine)—(yes)—(no)
(petrol aint energy)—(petrol in the tank)—(engines use energy)
(you have to burn it)—(fill a car and it's got energy)—(YES)
(makes heat)—(petrol has to explode)—(give out energy)
(works with air)—(it has to work something)
```

If we add to the evidence from these discussion trails the longer passage of explanation about nuclear energy quoted earlier on page 68, the full implication of these pupils' insistence on 'a source' of energy begins to become clear. It seems that energy, like energeticness, is an active and even violent thing. It has a lot in common with explosions and chemical reactions, but not much with the explosive substance. This may be the source of the explosion but is not itself either the explosion or its energy. Perhaps the analogy is that a tap may be the source of water without being water. The willingness of these pupils to discuss energy and the breadth of general knowledge that they shared, eventually revealed an underlying meaning for the word which was fundamentally at odds with the nature of the physics concept.

The discussions left little doubt that this 'explosion' view of energy was widely held. Since socially acquired knowledge is able to accommodate different meanings in different contexts it was necessary to seek and fix it. The same sample (N=128) of Fourth Year pupils were now asked for a written answer to the question:

Evidence from Talking

Which of the following sentences is/are correct?
A. Food already has energy in it before it is eaten.
B. Food only has energy in it when it is eaten and is inside your body.

In the middle ability groups only 45 per cent of the pupils chose sentence A. In the top ability group the score rose, but still to only 64 per cent. In the context of food, the 'store' view is rejected in favour of an 'explosion' or 'source' view, by a considerable number of pupils.

Those who held a notion of energy in which it could be stored and yet continue to exist, were not always easy to identify in discussion. Sometimes a single anonymous voice is heard saying 'No' to a proposition that energy is only present when it is evidenced by activity. Contradiction or refutation in face of a vocal class consensus needs considerable social confidence. There were just two occasions, in separate classes, when the idea was put clearly into appropriate words by a pupil who had, it seemed, both the confidence of well thought-out views, and a social licence to lead. On both occasions the intervention was effective; it brought about a collective pause in discussion which sounded like a resolution.

P1.	The sun stores energy.
P2.	So does a candle.
...Teacher.	What does 'stored energy' mean?
P3.	Stored for later use. *Not active.*
....	
...Teacher.	Can energy be stored?
P4.	Potential energy is stored.
(P5.	Can you actually explain what potential energy is?
Teacher.	Hang on a moment.)
P6.	It is not really energy...until it has been released.
P7.	It is — in a *passive* state.

Change and Development as an Internal Process

Understanding how socially acquired knowledge is changed, as it might have been by the last two closing comments, requires consideration of how the two pupils — P3 in the first extract and P7 in the second — may themselves have made such progress, and also how a social consensus can be moved.

Classical theories of cognitive development, such as the genetic epistemology of Jean Piaget, assume an internal construction of knowledge through personal intellectual growth, which feeds upon observations and operations in the outside world. The children try to assimilate their experiences into existing thought patterns. As they reflect, dissatisfaction with the fit of the new evidence may set in. Then the children may begin to 'regulate' their ways of thinking and so reach higher stages of intellectual accommodation.

Piaget agreed that social and cultural forces took a hand in this regulation of children's thought through the medium of language.

> Without language the operations would not be regulated by interpersonal exchange and cooperation.... It is in this dual sense of symbolic condensation (abstract definition) and social regulation that language is indispensable to the elaboration of thought. (Biologie et Connaissance, 1967)

The effects of abstract definition will be examined in a later chapter; at this point it is the social functioning of language which is the more important. Piaget seemed to believe that social influences between children only operated when they were at the same stage 'with no element of authority or prestige' (Piaget, 1952) so that they would be conversing on the same cognitive level. This assumes that even social interaction between people must be understood in terms of the internal structure of their knowledge, and was characteristic of Piaget's highly logical position. He held that external authority, like that of a teacher or another pupil who is much respected, produces no more than rote learning, not cognitive growth. Indeed, Piaget was at pains to point out that his was a theory of epistemology rather than of child psychology. While it would be unfair to suggest that Piaget was not interested in children, his real goal was certainly elsewhere. He was searching, he said, for an understanding of knowledge itself. Piaget wrote of his developmental epistemology:

> (it) attempts to arrive at an understanding of the mechanisms of knowledge through studying their origins and development. (*Psychology and Epistemology*, 1972:15)

Now it is quite possible to use this epistemological approach to the energy data in the last section. The scores of able and less able pupils (as assessed by the school) on the sentences A and B, about the energy stored in food, suggests that some rather taxing cognitive feat may be required. The resolving effect of comments in the trails about 'passive' or 'not active' aspects of energy adds substance to this suggestion. It seems like social regulation. Perhaps this evidence points to a growing ability, on the part of these 14–15 year olds, to separate the *action* of the energy from its *potentiality* for action.

Once the solution to the 'energy source' problem is put in this logical form it can be fitted quite easily into Piaget's framework. Despite the emphasis which is often placed on logical combinatorial mechanisms Piaget wrote of the formal stage of thinking that it was precisely this ability to think about possibilities, and to subordinate perceived actions to such *possibilities for action* that best describe this high level of cognitive development (Inhelder and Piaget, 1958). For pupils P3 and P7 then, formal thought has triumphed and solved the dilemma.

Changes in a Social World

Piaget's is not the only theoretical interpretation of what happened in these discussions, nor does it take all the dimensions of the pupils' work into account. Vygotsky would certainly have placed greater stress on ideas carried over on the back of language. He wrote that 'the child's intellectual growth is contingent on his mastering the social means of thought, that is language' (1976). He changed the focus of exploration from the internal maturation of logical mechanisms to social interpersonal events, like discourse, that can also stimulate intellectual advance.

More recently Doise and Mugny (1984) have built upon both these traditions. They maintain that even talk with those who are less cognitively advanced can stimulate intellectual growth. They write in Piagetian terms about this discourse giving rise to internal conflict. 'Learning from the mistakes of others' can arise through the stimulus given to reorganize and restructure their own ideas.

Clearly that approach well describes the advances in formulating ideas about energy reported above: there were some very straightforward examples of learning through the use of language. In one of the discussions a pupil, who has just given an off-the-cuff opinion that foods only have energy when you eat them, can be heard changing her mind by reflecting aloud on another piece of knowledge probably picked up from the legend on a food packet.

'They say "energy-rich foods" so there must be energy in them. Oh, I'm wrong.'

But there are features of the social construction of knowledge which none of these authors address. If the meanings and notions in the social stock of knowledge need to be continually reinforced, then social consensus becomes of the highest importance. Consensus-building is a process which might completely by-pass cognitive structures. Agreement, and not precocious cognitive development, would be the more valued commodity. Indeed, there would be little point in intellectual advancement if it made taking part in ordinary social converse, with its medley of context-bound and inconsistent meanings, almost impossible.

Although the drive towards consensus is readily apparent in most of the class discussions, the knowledge produced is neither monolithic nor static. Both within a single discussion and over longer periods of time it can be shown to change quite markedly even when no formal teaching has taken place. The cluster of different meanings present in social knowledge probably enables some movement in discourse. Other changes may be led by particular individuals.

> One of the Fourth Year classes recorded above had also been recorded two years earlier. At this time they were asked if a piece of bread, lying on the bench, had any energy in it. One of the pupils, let us call him Mark, answered that it did not because 'it couldn't jump around.' This energeticness notion won instant approval from most of the others. One pupil, Errol, can just be heard trying to disagree, but his view is ignored and his voice is drowned out by the rest of the class making jokes about jumping beans.
>
> Two years later the same question came up again in class discussion. At once Mark began to stake out his previous claim that food could not have energy until it was eaten, but this time it was his view that was ignored. Errol offered the idea that energy was stored in the food and the others agreed with him. One week later, in the next brief class discussion on energy, Mark could be heard to have changed his opinion and have come into line with the rest.

That anecdote gives direct evidence of the dynamic of class consensus. Whether Mark had achieved understanding, or had merely moved in order to 'stand in with the crowd', is impossible to tell. Possibly the question does not even have real validity in the context of the social construction of knowledge where several concurrent meanings exist and local communication is a major objective. In the social world of schools, consensus can be hard on the mavericks — on either a disregarded Errol in the Second Year, or an abandoned Mark in the Fourth. Both will feel the pressure to conform.

In the Longer Term

The literature on social influence within groups includes factors like social control, the need to reduce uncertainty, and aiming at consensus. Social control is unequally distributed, some figures — like 'Mark the joker' — have more influence on particular occasions. Confrontation and refutation have curious effects. Even if there is only one deviant in the group, as in our example, he or she can generate doubt. Classic psychology experiments in which groups had to decide whether a slide was blue or green showed that even those who had appeared to be sure of their opinion would often look again to see if the contradiction could be supported. In the subsequent test some of these adult subjects then changed their judgement from blue to blue-green (Ashe, 1950).

Evidence from Talking

So far the time-scales of remembering or reflecting have been neglected. Both of these have the power to change ideas. If the effects of individuals' influence and reputation are taken in conjunction with the participant's later reflection on what was meant, then the immediate forces of consensus no longer imply unchanging viewpoints. Development and change can become natural features of the social construction of meaning over time. The data from children's talk gives little more than a snapshot of the continuing processes of reconstruction and its inherent potentiality for growth. They provide an invaluable interpretation of present meanings but offer few clues to how or when they may be added to or may change emphasis.

> Thus, throughout a conversation, speaker-hearers do not only 'speak' and 'hear', but they also construct a cumulative and idiosyncratic account of what has been going on. This account is a construction in that many events are excluded from it, and it is made of interpretations of events, rather than the events themselves. This construction...does not necessarily end when the talking stops and the conversers separate: it continues when participants reflect on what was said.... But even this is not fixed: the reflective meaning is always open to change because of new information available, or new insights achieved by the speaker-hearer as he reflects on events that are past, or talks about them to others. (Barnes and Todd, 1977)

Chapter 5

Learning as an Extension of General Knowing

The Need to Learn

Evidence from the last two chapters shows that the social stock of knowledge is more like a free-choice cafeteria than a ladder to a hierarchy of stages. Our children can and do help themselves to new meanings and new knowledge according to social opportunity, personal preferences and intellectual tastes. All of these probably change with age, so knowledge both alters and develops: but learning is a much more intentional process. This chapter will look at a variety of different ways of learning about energy, for both children and adults.

In an unschooled society further and wider experience of everyday life simply adds to the stock of knowledge without changing or belittling the value of what was there before. But in our society the complexity of modern living, with its burden of technical and inherited knowledge, makes schooling seem essential. There is a lot to learn and the haphazard selection of alternative meanings is a slow and inefficient way of acquiring knowledge. We also expect our children to have knowledge of what they have not yet experienced and indeed may never encounter. In our society knowledge is not just the outcome of what has happened to us in the past, it is also the key which may open doors to valuable new experiences, for example through the job market, in the future. It is widely believed that, with so much knowledge to be mastered, children are in urgent need of a more forced pace of learning, so we pack them off to school for compulsory instruction.

With few exceptions the topic of energy figures in this school curriculum for one or more of the following three purposes:

Energy education for the citizen.
Energy education for vocational training.
Energy education for acquiring scientific knowledge.

Learning as an Extension of General Knowing

The first of these begins with general knowledge and leads on towards the kind of action that citizens need to take to regulate their own 'consumption' of energy, and to comment on public energy policy. It will have relevance both to the household fuel bills and even, on occasion, to the ballot box. The second purpose for energy education begins with the skills and knowledge associated with doing and making; it leads on towards the technical know-how needed for employment or private hobbies.

The third goal of energy education is academic. This purpose, seemingly unrelated to everyday living, makes it substantially different from the other two, and implies that this type of education may not follow on smoothly from socially acquired knowledge. Indeed, the characteristics of scientific knowledge contrast so sharply with those of common-sense knowing, that this kind of education presents special problems and will be treated separately (Chapter 6.) Following a course in school on science, and profiting from it, is not synonymous with acquiring true scientific knowledge. This is the type of learning to be considered here.

Embedded Knowledge

The common-sense attitude, that justificatory system behind the barely connected stock of meanings and information which we call general knowledge, allows for plenty of diversity. Being weak on overall logical structure it permits all kinds of additions. In the last two chapters, when examining what the pupils had written about energy, or listening to how they spoke to each other about what energy means, this vagueness was an obstacle to outside investigation. Indeed the term 'meaning theme' had to be invented to describe what the pupils seemed to be saying, since 'theory', which suggests consistency and structure, seemed inappropriate. But for informal learning, in which new knowledge just finds a niche amongst other meanings, this lack of structure may become a distinct advantage to the learner.

Children's untutored thinking can be powerful. Mary Donaldson (1978) has written eloquently about primary school children, and the difficulty they find in solving abstract theoretical problems, compared with their deep skill with the same kind of problem when set in a real context that can be readily visualized and understood. She called the two kinds of thinking 'embedded' when it was in a familiar context and 'disembedded' when it was not. Embedded knowledge, she wrote, was available to 'human sense' — by which she may have meant much the same as our common sense — in that it is socially shared and situationally bound. She drew attention to the way in which even the youngest children can respond socially to the intentions of the adults around them.

By the time they come to school all normal children can show skill as thinkers and language users to a degree that must command our respect, so long as they are dealing with 'real life' meaningful situations in which they have purposes and intentions and in which they can recognize and respond to purposes and intentions in others. These human intentions are the matrix in which the child's thinking is embedded. (121)

The reference to the social skills of children, and to the purpose and intention of the learning is useful and matches well with life-world knowing. However, the word 'embedded' suggests that children will not be able to abstract general ideas from the concrete instances they have met in 'real life'. This is not quite the case. We have already heard untutored secondary school pupils giving definitions of energy as 'What makes things move', and imagining it as a stored and passive entity. This is a kind of disembedded knowledge, but it remains different from the abstract knowledge of science if it is just added to the general stock of knowledge alongside other examples. The purpose and intention of disembedded knowledge is not abstraction for its own sake. Rather it is that the conceptual term, or mathematical formulation, can encompass many different contexts. If abstract ideas, like energy as the potentiality for action, are used without any effort to build them into a system which might overarch a range of everyday energy contexts, then it will not be a radically new kind of thinking.

There is little doubt that many of our secondary school pupils react to their science teaching in just this way. Sometimes when they say, 'That's just general knowledge', it shows that the lesson seemed easy, perhaps even boring. What it always indicates is that the internal structure of the new scientific knowledge has not been recognized, but only the bits and pieces of the information it provides.

Is Physics Easy?

This informal way of learning can be illustrated through the words of a class of average ability pupils, aged between 14 and 15, who had just begun a teaching unit on mechanics in their physics lessons, which included the topic of energy. They had been taught some of the simple theories and formulae, such as the transference of energy from one system to another and the transformation from one kind of energy to another. Because they had initial difficulty in using and understanding the vocabulary, much of the introductory work had been carried out orally through questions and answers and class discussion. Then they graduated to experiments and written physics exercises of the normal kind and seemed to achieve some success. However, when the class were given a written physics test at the end of the term most of them did not acquit themselves with any great distinction. Each pupil was interviewed separately after the test by the teacher to discuss where they had gone wrong, and to be helped to put it right.

Learning as an Extension of General Knowing

At the start of each interview the pupils were asked if they had found physics difficult. Out of the seventeen pupils interviewed thirteen said they found physics easy, adding that there was not much new to learn, that they had already known about force and energy, or that it was 'just common sense', or 'just general knowledge'.

P.	Physics is easy because it's about what you know.
Teacher.	You reckon you didn't learn much that was new?
P.	No
Teacher.	No? How did you know it already?
P.	Acceleration, I knew what acceleration meant. And about energy and all that.

Now this pupil had scored very low marks on the test paper, and got the whole question on acceleration completely wrong just because everyday meanings, not physics definitions, had been used throughout. The teacher's problem was to get across the idea that although the words used in the physics lessons were the same as in daily talk, the meanings were not.

Teacher.	Is work something you knew about already?
P.	Yes
Teacher.	Is the word 'work' used in the same way in physics as it is in everyday speech?
P.	Yes because, say, when you get a job or something, that's one kind of work. But another type of work is when you hold something up.

It is the hallmark of general knowledge learning that no distinction is drawn between one kind of meaning or explanation, and another. As the younger pupils had remarked about the energy/exercise contradiction (page 52) 'it all depends on the situation'. This pupil had received the physics definition, misremembered it, and put it on a par with everyday alternative meanings.

The pupil was a particularly poor achiever and even the example of physics learning that he gave was incorrect. The next extract comes from another pupil who had worked hard and added some useful items to an existing stock of general knowledge. It is particularly interesting to hear him comment that the teacher had expected an extension of the more usual run of 'energeticness' meanings. He was aware of having learnt the notion of energy as the potentiality for action or work. His social skills, as Mary Donaldson remarked, are impressive.

P.	Physics seems easier because it's down to common sense a lot, you know.
Teacher.	When you say common sense, some of it you might even have known before?

Getting to Know about Energy in School and Society

P.	Actually it's because it's to do with everyday life, a lot of these things, aren't they? It's very practical, like. It's a very practical subject.
Teacher.	I'm interested you say that. Look at that question, for example. (Question asks what 'energy' means.) You've put down 'Energy is the possibility of doing work'. Would you have known that before you did the course?
P.	Well, I probably would have known it from those terms, but I didn't kow how to put it into such a precise manner.
Teacher.	How would you have used the word energy before?
P.	Energy isn't doing work is it? But that's what you term it as. That's what I would term it as. Because even in everyday language you say things like 'energetic' and so you imagine someone actually running, or actually jumping, or someone actually kicking a football. So you imagine they are actually doing work, not the possibility of doing work.

That pupil's answer to the test question which asked him to name some kinds of energy finds him writing 'a battery', and 'water behind a dam'. Both of these fit in well with the stored 'possibility' of doing work which he was conscious of having picked up from the course. In this sense his learning had been satisfactory: it amounted to a real extension of his general knowledge. The pupil could pick out examples in which energy was stored up: as he commented in the interview, 'I knew what you wanted...'. Like the younger children that Mary Donaldson had described, this 14 year old was competent at sensing adult expectations although not always capable of fulfilling them.

In the sense of orthodox physics knowledge, however, the answers he gave were incorrect, and not what the teacher had hoped for. It was good to know that the pupil had enjoyed the course and found it satisfyingly practical, but the primary aim had been to teach an abstract terminology which could be exemplified in many practical situations. For this pupil only the situational examples were accessible during the test. The 'kinds of energy' required by the test question should have been disembedded from any situation as concrete as a battery or a dam: it should have been stated as *electrical energy* and *potential energy*, as he had been taught to do during the course. Instead of this, the new knowledge had been received as yet more general knowledge and added to the general stock.

Inevitably, the marks scored by this pupil in the physics test were not good; but perhaps the education, as he received it, was. Something in the manner of its teaching or its learning had made the new knowledge acceptable to him, not as a difficult alternative view of the world, but as a nonthreatening elaboration of valuable familiar knowing. He saw its purpose as 'to do with everyday life', and that made it easy, as Mary Donaldson had suggested.

Citizen Understanding of Energy

What kind of knowledge about energy does a citizen need, and for what purposes? Not, one would assume, for answering quiz type questions which often figure on examination papers, such as 'Which of the following units measure energy.........'. Nor is it likely to be an understanding of theories or conceptual terms. It can do little for the plight of citizens to know exactly what 'kinetic energy' and 'potential energy' mean, and in chapter 8 we shall see that a knowledge of the Conservation Principle can sometimes do more harm than good for learning about the generation and use of energy.

Scientific knowledge suitable for adult living will illuminate the concerns and values which press upon adults. Although Shen (1976), Miller (1979) and others have written about *cultural scientific literacy*, by which they meant an understanding of the structure and indeed 'majesty' of scientific thought, that view seems almost out of date from the perspective of the 1990s. Global warming, food contamination, and water pollution receive anxious citizen attention not because they represent any pinnacle of theoretical modelling but because they threaten to impinge upon daily living. More access to information, not knowledge about the scientific culture, is the popular call.

That may seem too summary a dismissal of public interest in science for its own sake. Indeed, the relative popularity of books such as Stephen Hawking's *A Brief History of Time* shows that it does indeed persist. The history of the dedicated amateurs in science has been well explored by Layton (1987). Scientific and literary clubs flourished in the nineteenth century and certainly numbered amongst their members some who made real contributions to scientific thought. Hermann von Helmholtz, who published the first unequivocal statement of the Principle of Conservation of Energy (*see* chapter 9), was himself a self-taught amateur in respect to his impressive knowledge of physics. For such dedicated amateurs — they can hardly be called a 'public' — the appetite for science is such that they are quite prepared to absorb the abstract discipline of scientific thought. They certainly would not have added this hard-won scientific knowledge to their general stock of knowlege, as though it were structureless pieces of information. For them science was an end in itself, not a response to personal crisis or risk in the circumstances of daily living.

Today it seems that science affects life far more than ever before. That cliché may well be untrue in almost every respect, except in the public rhetoric. But that is precisely where it has the most powerful driving effect on the average adult's urge to learn. They are being told continuously that their quality of life is being adversely affected by science and technology, and it is out of this anxiety they seek more information.

Adults' new information is likely to be as much bolt-on bits and pieces as the pupils' half-learnt knowledge discussed in the last section, but in this out-of-school context it is not a deficit state, and cannot be stigmatized as either misconceptions or ignorance. Adults may badly need extra information and if the addition sits easily with their mature understanding, with knowledge and concerns from their work place and their homes, it will have real value as citizen knowledge.

There may be arguments and problems about energy to be considered in terms of social justice within the community. There will also be personal values and public standards related to how we want to be seen using energy by our neighbours, such as entertaining in the 'front room', or using the green hose at the petrol station. Everyday morality, and the maxims in which it is embedded, form an integral part of the social stock of knowledge. Being as structureless as common proverbs and common knowledge, they are not to be confused with the discipline of ethics. Indeed phrases like 'wasting energy' carry practical knowledge and moral disapproval in almost equal and inextricable senses. Physicists may deplore the content of such phrases, but for public understanding they can have great advantages. Scientific information needs to be partnered with complementary social understandings, even at the expense of conceptual purity, if it is to become usable as citizen knowledge. This theme is further developed in chapter 8.

Barriers to Adult Knowing

Thinking like a school child, even in scientific terms, will not match well with the familiar responsibilities of adult life. By this criterion the following passage, from a 50-year-old woman, seems sadly out-of-step with her everyday competence.

> 'You know, you take in what happens around you now, but I think you've got a basic thing in you that...you know, your education as a child — what you learnt then. And I think it's always there and it comes out. And you think on the basis of that, if you know what I mean. When I was little there was no such thing as solar heating. Well, nobody talked about it. It may have been thought of by scientific people, I don't know. But I find it difficult to understand it fully. Nobody has ever bothered to explain it fully to people like me, so you just pick up bits as you go along.'
> 'Well you obviously have done so.'
> 'Well it is interesting, but I never have the time — or really that much interest — to sit down and read about it.'

It was reported by Irwin and Jubb (1989) in their study of people living near a large chemical complex, that many could talk well about the range of risks they would accept in relation to employment and housing, despite a lack of chemical knowledge. The greater part of the research was carried out by informal home interviews with the context set in a familiar adult world. When (rarely) the interviewer switched to factual school-type questions about science, she reported a complete change in attitude, mannerisms and confidence. Respondents reached nervously for their glasses, or stammered. The context transformed the questioning into a judgmental test, and demoted the skills of the adult knower to the crudities of a scale of factual correctness which they probably had not met since their school days.

A lack of success in school science can also build up an awe for its supposedly arcane structure so huge that it forbids all subsequent entry. This not only debars the citizen from access to valuable information; it can also be carried over from one generation to the next. Studies of primary children who have not yet begun to learn science often report them saying 'You have to be brainy to do science — *I couldn't do science*'. Any suspicion that there is a logical structure to scientific knowledge only serves to heighten this psychological barrier. The following extract from adult interviews shows a graduate English teacher using terms heavy with obscurantism to explain her ignorance of scientific knowledge about energy.

> 'It could be said by scientists, *who know about these things*, that there were certain ways in which energy can be created, transformed, used, um,...that there are limits and parameters within which this can be done. I don't know anything more about it'.

Public understanding of energy needs to be embedded it seems, in mature social needs and purposes, and related to daily living. Only then can it have the dignity of confident adult knowing.

About the Processes of Science

Amongst the information about science which the public picks up there may be some items about its methods and procedures. Although it is now commonplace in education to refer to these as 'scientific processes' we should be just as chary of ascribing to them a universal character, as we were of describing the common-sense knowledge of science as theoretical. Statistics gathered from surveys on the *Public Understanding of Science* (Durant, Evans and Thomas, 1990) emphasize this point. The authors asked the same 'process' question in two different scientific contexts during a European survey. Respondents were asked which method would be used to tackle a scientific problem. In the first question the context was medical: which of two medicines should be used for treating blood pressure. The

Getting to Know about Energy in School and Society

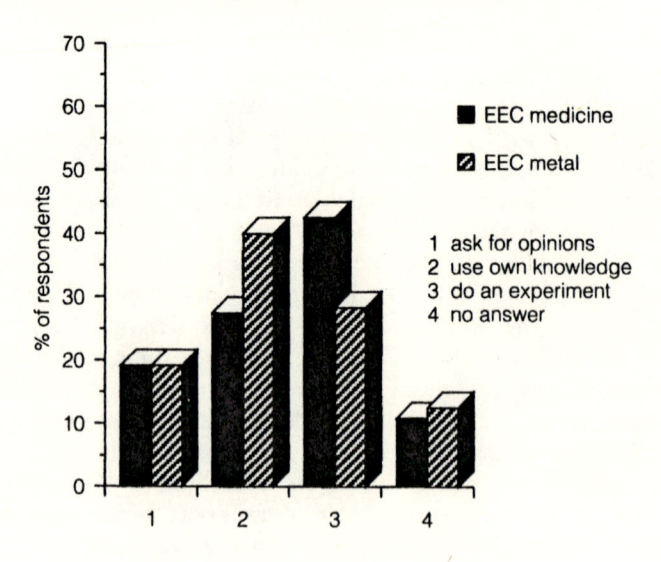

Bar chart showing % of respondents for EEC medicine and EEC metal across categories: 1 ask for opinions, 2 use own knowledge, 3 do an experiment, 4 no answer.

second question used an example about deciding between two metals for a special use. The graph above, from their paper, shows how very different the responses were to these apparently similar questions.

The authors themselves interpreted this anomalous result by concluding that 'medicine is the paradigm for science' in the public consciousness. By this they seemed to mean that medicine was somehow seen to be more scientific, or was the paramount example of how scientific research should be done, in popular esteem. It would be just as simple to assume that there is only context-dependent understanding of scientific process.

School children's understanding of science also casts a different light on these results. Asking 11–14-year-old pupils about how they think experiments are done reveals that about half of them believe scientists have no expectation about what the experimental result will be before they carry it out. The scientific process is thus reduced to little more than a 'shot in the dark' affair, adding yet more information to the atheoretic jumble of knowledge that scientists possess. Durant and his co-workers found a rather similar result when they asked directly 'What does it mean to study something scientifically?'

Science as theory building or hypothesis testing.	3%
Science as experimentation without reference to theory.	11%
Science as fact-gathering by use of microscopes etc.	43%
Don't know, no answer.	43%

These snippets of evidence suggest that the scientific process is something of a closed book to both school child and general public. It is thought to be context-dependent; experiment, its chief glory, appears to be unrelated to theory as an overarching predictive knowledge structure.

For those like Millar (1989), and indeed the *ad hoc* group of the Royal Society that reported on the Public Understanding of Science (Bdomer 1985), these results must be worrying. They had put knowledge of 'the scientific process' high on their shopping list for public understanding of science. From a perspective in the sociology of knowledge, however, these findings simply confirm that citizen understanding of scientific methods is also accreted to general knowledge. As such it declines to erect any universalistic structures, and this permits citizens' responses about process to change complexion quite radically as their knowledge is adapted for use in other situations and for other purposes.

Energy Knowledge for Survival

The average citizen rarely has to answer survey questions about energy, but few can avoid operating in a way which requires some knowledge about it. To judge from recent work on the energy budgeting of the elderly (Jenkins, 1990), it seems that people manage this in ways which are quite surprisingly rich in diversities of personal style. Far from just 'learning' a new body of knowledge for regulating their behaviour, old people in this study were shown to impose their own values of economy, frugality, keeping up appearances, and realism about life-expectancy, to guide both their use of energy and their understanding of it.

Quite the most surprising outcome of most recent studies of the public's understanding of science is people's occasional *rejection* of knowledge about life-threatening situations. If understanding more about energy had real survival value — the promise of prolonging life or avoiding great risk — it might seem immediately obvious that people would be strongly motivated to learn. Curiously enough the evidence does not always confirm this commonsense expectation. In Jenkins' study the elderly people had thermometers installed in their rooms by the social services so that they could see when the temperature dropped to a point where they might be at risk from hyperthermia. Time after time Jenkins reported that the old people admitted that they did not bother with the thermometer. 'I never look at it', or 'If I hadn't got one I wouldn't bother to get one.'

This observation may have a sociological basis far deeper than just the entrenched obstinacy of the aged. Arguing about other people's reactions to risk calls for a very delicate understanding of *what people can allow themselves to contemplate*. Risk perception in general, and in relation to nuclear power in particular, will be considered in chapter 8. In the context of daily living, hazards and dangers are what we all strive to avoid if we possibly can. The human race has survived through this reaction. In our complex society it is easy to see that avoiding danger now requires knowledge just as much as fast reaction-times. It also involves taking action. The citizen expects to be active in warding off risk, so that risk avoidance becomes a part of the way of living. From such a perspective people who are not able to take action to avoid risk cannot allow themselves even to think about it. So the thermometer that warns the elderly to turn up heating, which they may not be able to afford, needs to be ignored.

Brian Wynne came across a similar and curiously deliberate obliviousness to survival knowledge amongst the apprentices at Sellafield. He wrote that they:

> vigorously defended their scientific ignorance, when we had assumed that such understanding (an elementary knowledge of radioactivity) would be crucial for their own safety. (Wynne, 1990)

A similar angry rejection of knowledge was recorded by McGill (1988) in her study of the Sellafield workers' reaction to a television team who wanted to monitor the radioactivity in their homes. Not only did they refuse access; they subsequently ostracized the one member of the community — the local postmaster who was not employed at the Sellafield works — who did provide the contents of his vacuum cleaner for monitoring.

Just as the old people in Jenkins' study did not possess the economic resources to act upon information about hyperthermia, so the workers and apprentices at Sellafield had no way to alter the safety procedures built into their place of work. All three groups were effectively deprived of the option of acting as 'risk avoiders'. So, rationally enough, they studiously avoided all knowledge about the risk.

Generally, one suspects, this avoidance of knowledge is not made explicit and the ignorance is comfortable — if not actually bliss. However, should this lack of knowledge, with its concomitant lack of power to act, be consciously reflected upon, it can produce an understandable feeling of angry alienation. A comment of this kind was recorded in one of the initial adult interviews about energy.

> 'Can I just say that when we are using coal, the simplest form of energy, most people understand it. It's acceptable. But nowadays who understands nuclear energy? Now we've got to the position when only a few experts — physicists, you know — understand it. So we've lost power here, you know. We are powerless.'

Energy Knowledge for Technical Purposes

Both school children and apprentices will learn much of their knowledge about energy in this way. Keith Ross (1989) established that firemen had an untaught understanding of heat and burning which was considerably greater than most other groups in the community. This must have been the result of learning for a purpose: its survival value is easy to see. Other technical knowledge about energy may have less obvious life-and-death value, but it is learnt both for a job, and very largely on the job. That also affects its character.

The educational literature is thick with definitions of technology and it would be an impossible and pointless task to choose between them. However, all are agreed about the paramount position of 'purpose' in any technological exercise, and most state this in terms of 'human needs'. In an industrial setting it is not always so clear whether the 'need' is that of the manufacturing firm or of the recipient. Be that as it may, it is the purpose and intention of the work which embeds it so clearly in the world. As Donaldson's thesis predicts this makes its learning not only easier, but is another way of adding to general knowledge.

The school technology syllabus usually includes a unit about energy. It will be different to a science course on energy not only because of its different purpose but because pupils recognize and react differently to the context of the workshop. I taught an energy unit for the CDT (Craft, Design and Technology) department of a school twice during one academic year. The pupils began with lengths of suitable wire and were required to find out the best way to coil it in order to make a heating element. A Joulemeter was available to measure the energy flow and instruction was given on its use, but not on any theory. As the children tried winding coils of different diameter and density of turns they gradually got the coils to glow red hot, at least in places. Success was 100 per cent and so, it seemed, was satisfaction. The results were quite spectacular but, on each occasion, only two out of the 30 pupils asked me 'why' the coils got so much hotter when they were wound close together. The rest absorbed the new knowledge as an unproblematic 'fact', not needing explanation. It was added to their general stock of knowledge and used to devise other electrical heating gadgets.

Procedural Knowledge

Another kind of knowledge, unthinkingly related to action, controls our movements. Its purpose is also crystal clear. It is embedded in the mental programmes which allow us to talk about other things while opening a door, playing the piano, or riding a bicycle. The complexity of its apparent simplicity has provided headaches for cognitive scientists, computer theorists, philosophers, and educationalists alike.

Piaget's theory of genetic epistemology is based upon the relation between knowledge and action. As children meet active experiences in the natural world they begin to understand them by a process of assimilation to their own existing ways of thought. But Piaget encountered some cases

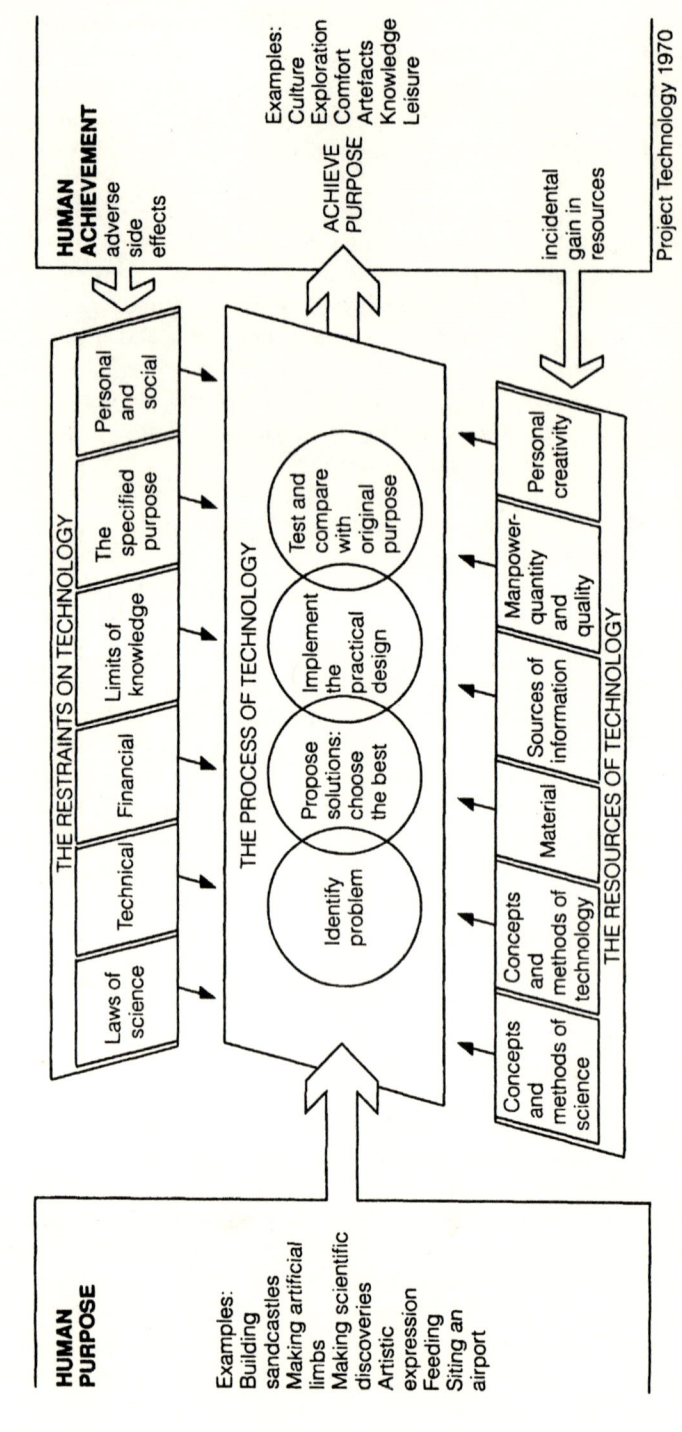

Getting to Know about Energy in School and Society

Technology — A disciplined process using scientific, material and human resources to achieve human purpose.

Project Technology 1970

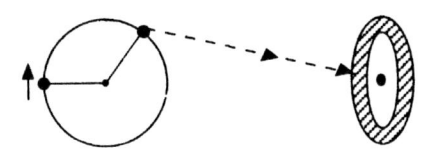

Piaget's task with sling and pellet

where it seemed that thinking was not ahead of action. There was one test in particular which could not easily be fitted into his original scheme (Piaget, 1976). He set his child subjects a task where they had to aim a pellet at a target by releasing it from a rotating sling. Most of the children were able to do this successfully either straight away or after a little practice. The trick was to let the pellet go while at the top of its trajectory so that the tangent to its circular path aimed directly at the target. Obviously, if it were released at the nearest point to the target it would travel at right angles to the required direction.

A curious finding of this research was that a large number of children who managed to strike the target, insisted afterwards that they had released the pellet from the sling when it was at its closest to the target. It seemed that the children could carry out the task better than they could think — or at least speak — about it. Piaget wrote that they 'could perform precociously successful actions'. This is similar to McCloskey's findings of understanding the physics of dropping a ball into a hole (page 45).

This practical knowing, like diSessa's p-prims mentioned in chapter 2, rarely gives rise to the kind of successful learning about phenomena that might have been expected by cognitive psychologists. School physics teachers are also often aware of children who have motor knowledge of Newton's Laws of Motion — such as subconsciously bracing themselves for force during the acceleration of a bus — without the associated articulated knowledge of the concepts. My own experiences in school physics teaching (which included a final exasperated trundling of a heavy 15-year-old pupil round the laboratory on a kitchen trolley!) confirmed that this knowledge was almost always inaccessible to the pupil.

Making Tacit Knowledge Available for Learning

The results of the 'category 6' studies of the Assessment of Performance Unit (1983, 1984) provided more evidence of this unspoken knowledge. Again there were cases where the pupils carried out better investigations than the accounts they gave of them. The observers reported on the controls and measurements that were actually made. At the end of the tests they asked each child, 'What did you do to ensure that it was a fair test?'. In almost every case the children gave themselves credit for far less in the way of experimental controls than they had actually used. The written conclusion in the published reports was that the children 'knew more than they could say'. It seemed that the practical situation had elicited tacit procedural knowledge of how to carry out a test which the children could not put into words.

According to Michael Polanyi (1958) there are three categories of tacit knowledge. In the first of these the knowledge is totally ineffable. He illustrated this by reference to riding a bicycle, or recognizing one's own raincoat. In both cases the action can be performed without articulation and it is hard to see what articulation of the relevant knowledge could add to action. This is not a learning situation in the sense of added knowledge.

In the second kind of tacit knowledge we articulate one aspect of our work at the same time as using a related tacit component. When we read, for example, we attend to the meaning of the message while tacitly using the familiar symbolism of the written words which we learnt so long ago. The experienced typist or craftsman uses tacit skills to extend their range of perception beyond their finger contact with the instrument they use. Polanyi gives more illustrations of this curious personalized tacit skill by which we move into the domain beyond the tool, as though it has become an extension of the body. It is how drivers adjust the range of their body feeling to that of the chassis of the car itself and even wince if they drive too close to a wall! The focus of attention moves outwards from the limits of the body, to where the car, or the familiar instrument being controlled, makes contact with the outside world. It is what Polanyi calls 'in-dwelling' — the final achievement of learning to become at one with the tools being used.

Conscious learning begins where this tacit knowledge ends. In experiment the thermometer becomes an extension of the probing finger, and then understanding of temperature begins. Similarly the APU children could only tell the tester about plans that they had consciously made and carried out. What they did 'without thinking' escaped articulation because it was too close and too familiar.

In the third type of tacit knowledge the two kinds of knowing have 'fallen apart', as Polanyi says. This means that each can be used to explore or regulate the other in a very valuable way. If articulate knowledge runs ahead, as when we speculate out aloud, then tacit knowledge may be used to check it. 'That doesn't feel right', we say.

Tacit knowledge acquired from action is strongly situational, and 'taken for granted', much as general knowledge is. But it is different from general knowledge because it is almost exclusively personal and cannot be reinforced or added to by social converse until it has become verbalized. If we wish to produce understanding about energy using tacit knowledge, it seems that we shall have to give our children words to use, so that they can bring their tacit knowledge to bear on how things work, and so relate this to experiments and investigations about energy in the school laboratory. This brings the argument back to Vygotsky's useful vision of words as the necessary tools that children need to probe experiences.

Learning from Group Practical Work

Practical work has been a great focus of attention recently (Woolnough, 1991), yet it is curious that, amid all the talk of 'explorations', 'assessing skills', or 'holistic tasks', the use of pupils' own laboratory experiences for learning about concepts, as opposed to demonstrations and questioning, has been very largely ignored. I have written elsewhere (1980) about some aspects of this problem. In particular I tried to show that the effects of theory-laden observation implied that pupils could *not* learn accepted concepts or theories on their own, by doing and discovering, because they would be likely to interpret what they saw in line with their own existing preconceptions. This is a point which relates to the constructivist view of teaching (to be discussed further in Chapter 7). Here the central point about practical work in the laboratory is that *conceptual understanding is not so much an outcome of experimental work as an essential prerequisite for its successful operation.*

The problems to be tackled in the teaching of energy through laboratory exercises are:

- how the experimental work can link to existing general knowledge about energy,
- how groups of children can use concepts to make a personal and social reconstruction of what they are doing, and
- how this can give speech to the pupils' tacit knowledge so that they can reflect upon action.

The first of these problems involves pin-pointing the knowledge suitable for the pupils in question using an understanding of what children at this age might know, expect, and feel. If we assume that the class is in the First or Second Year of secondary schooling (age about 11–13) research findings from chapters 3 and 4 would predict that they possess:

- a largely personal concept of energeticness, and
- some extension into the non-living realm, and
- only little progress towards a measurable and generalized concept, and that
- the notion of passive stored energy may be missing.

Starting from 'energeticness' suggests that laboratory experiences where the children can 'feel' their own energy at work would be just right. We shall need practical projects where the children set a machine going by transferring their own energy to it (winding up), where a weight is raised a measurable distance, and where the energy can be held or stored in a fairly obvious way by the device. I am indebted to my colleague Brian Woolnough, a collector of wonderful toys, for 'feeling' your way into physics, for an introduction to hand-driven dynamos which became heavy to turn as they were connected to an electric circuit. Energy could almost be felt flowing out of the body and being drawn through the dynamo into the electric circuit.

Before, during and after each of these exercises the teacher will be using phrases like 'transferring your own energy', 'Have you measured what the energy can do?', and 'Is there stored energy in it? — flowing through it?'

> ... the importance of language is that it makes knowledge and thought processes available to introspection and revision. (Barnes, 1976)

The thinking behind the practical exercises being planned may fail unless the knowledge is reinforced by social exchanges within the group. If the pupils are left without the necessary verbal and conceptual tools for talking about their experiences social construction cannot take place. The sort of chatting that goes on in all group practical work is necessary for the allocation of tasks — fetching the apparatus and lighting the bunsen burner. How it is said, the words used, and the responses of others, make a humble but essential contribution to the meaning of phenomena, and hence to pre-conceptual learning.

The aim of using the newly articulated knowledge to reflect on action being taken, as Polanyi advocated, can only be achieved if the pupils are allowed a real dimension of freedom. Practical work regimented by recipe-like worksheets does not call for reflective talking. No amount of lifting weights from the floor to the bench top for 30 seconds will do. The experimental work will need to be open-ended technological investigations of the type where the pupils might say 'I think it would be better if we...', or 'Let's try and see if...'. In this way it begins to fulfil Polanyi's criterion of using tacit knowledge during reflection on action and allows the emerging ideas about energy to become the sort of 'explore-withs' that Robin Hodgkin described in *Playing and Exploring* (1985).

Explorations about Energy

The practical exercises will be bound to have conceptual objectives such as the storage of 'passive' energy and energy transfer from person to machine. They may also challenge invention or competition. The aim of all this activity is to get the pupils to practise using these ideas, so that they make them their own by play and exploration. (As the extracts in the next section show it is not always possible to tell where play ends and exploration begins.)

Learning as an Extension of General Knowing

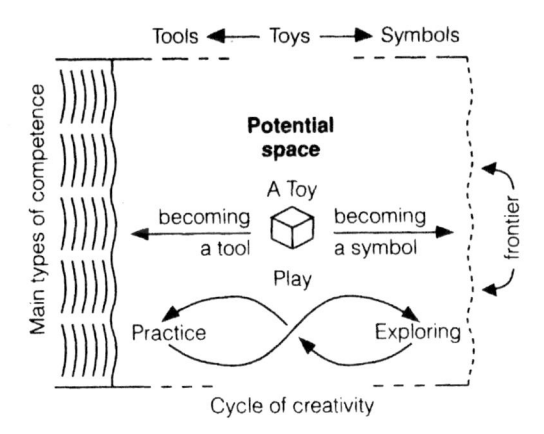

Diagram: Axes labeled "Main types of competence" (vertical) and "Tools ← Toys → Symbols" (horizontal). Central "Potential space" containing "A Toy" with arrows "becoming a tool" ← and becoming a symbol →. Below: "Play" with cycle between "Practice" and "Exploring" labeled "Cycle of creativity". Right side labeled "frontier".

From *Playing and Exploring* by Robin Hodgkin

Thirty Second Year children are at work. They have been asked to make a water turbine out of curved pieces of stiff plastic which they fix into a rotatable cork. They use this model to raise a weight. Their assignment includes questions like 'Should the blades be short and fat or long and thin?', 'Where should the water flow so as to give the most energy to your turbine?' and 'Measure how high your turbine can raise a weight in 5 seconds?'

> Steven and Matthew ignore the bowl and beaker they have been allotted and rush to stake a claim to the nearest sink. Triumph! I am called over to watch it work — but it doesn't. Steven is directing the water straight down on to the middle of the cork. I prompt Matthew 'Where should the water flow?' After a moment he changes to the edges of the blades and it spins. I praise Matthew and use the words 'energy' and 'leverage'. A little later I return, show more appreciation, and ask another question with the word 'leverage' again. By the end of the lesson they can be heard using it themselves while they alter the blade length.

It is not uncommon to find groups of girls who are not attracted to the competitive element in this sort of work. Given space they too can find ways to use their conceptual 'explore-withs'.

> Christine called me over. 'Our's is a complete failure', she said. The others nodded. A sticking axle meant that it only turned feebly. Anna said, 'It only lifts the weight while it is under the water'. I asked if things weighed less under water. They talked this over, and then asked if they might leave their model and look at the others. They stood well back from the models and talked together. A little later I found them drawing 'an ideal turbine'. It was good.

Activity can also be used in a public way to gain, or to retain a cherished social reputation.

> Danny's group were falling about at the back, as usual. They were making a 'potential energy machine' in which a wound-up weight operated a fan. I walked past and asked who was going to put energy into the model. 'Me' shouted two of them together. About twenty minutes later they shouted 'Look at this!' One of them had climbed up precariously on top of the back cupboard. 'This has really got potential energy!' said Danny. Once we were all watching they let go; I thought the whole cupboard was going to fall...but it didn't!

Of Additional Knowledge

This chapter has reviewed the different purposeful ways in which children and adults get to know more about energy without actually going through, as it were, the portals of scientific knowledge. That is not to say that the learning is without value or that it is misunderstood; quite the contrary.

West and Pines (1986) have given an analysis of children's learning which has superficial similarities with the one presented here. They have carefully distinguished two *sources* of children's knowledge — that which they acquire spontaneously from their interactions with natural events, and that which they are taught at school. (The rest of their argument, about the possible kinds of conflict between the two sources of knowledge, is interesting and will be considered later.) However, the argument in this chapter is *not* that the sources of knowledge are different, but that chunks of it may be differently received, stored and valued. Informal knowledge often contains items of correct taught science which are added on to the receivers' other items of knowledge, beliefs, and imaginings. Conceptual conflict simply does not exist for this kind of learning because it is evaluated in the manner of life-world knowings as useful for different occasions and permissive of contradiction.

Some children learn like this in school science lessons; adults learn this way when anxious about some technological threat, and pupils do it during practical investigations. In every case the learning process has odd features. Because it lies close to social knowing it is influenced by affective factors such as fear or feelings of status. Sometimes the knowers hardly know they are learning. The over-riding features of all this informal learning is its life-world purpose, and its lack of shape. Even, as we have seen, knowledge about scientific method seems to shift and lack consistency.

Tacit knowledge derived from doing is the most impenetrable of all. When Polanyi first approached this topic its chief relevance seemed to be to technical training. Now that 'explorations' have become so central to education in science that some educationalists such as Qualter *et al.* (1991) can claim that all science should or can be learnt this way, it needs far closer examination. In the interactive science centres which have sprung up like mushrooms during the last few years the debate about whether and what visitors are learning is just beginning (Russell, 1989, Solomon *et al.*, 1991.)

Chapter 6

The Formal Domain of Scientific Knowledge

Meeting Scientific Theory

If children, and adults too, learn scientific information almost painlessly when it is absorbed directly into the fund of general or life-world knowledge, why do they not succeed equally well in school science lessons? What makes formal scientific knowledge so much more difficult to learn? And does it have to be quite so different from common-sense knowing?

Some answers spring readily to mind. The purpose and need for information, which characterized general knowledge learning, is missing. It is not simple information, but a whole way of thinking which is to be striven for, and only a deep commitment to science can make that a personal goal. This removal of practical need as a motivation to learn, is caused by another difference: scientific knowledge is not related to any specific context. It is true that the abstract theories and concepts of physics demonstrate their power most clearly when they are used to solve practical problems; yet they themselves are undeniably removed from the objects of daily life. This enables them to be applied to all and every situation. These theories and concepts concern *force*, *energy*, *work* or *acceleration* in the abstract. That makes the knowledge 'disembedded' as Mary Donaldson wrote, and so harder to learn. Just how this difficulty shows itself will be the subject of this chapter.

The basic purpose of this science teaching is to instruct our pupils in accepted definitions and established theories; it becomes a kind of cultural transmission. That objective may sound almost brutal in its straight-laced intention of passing on certified knowledge, as unaltered as can be managed, to the next generation. To borrow the hoary phrase attributed to Newton, our pupils are being hoisted 'on to the shoulders of giants' so that they can see the view of nature that others have had, and then, perhaps, a very few of them will see further still.

The major difficulty, which is met again and again in the teaching of physics, is the leap from the life-world domain of affective reactions and alternative context-dependent meanings, to uniquely correct definitions of

The Formal Domain of Scientific Knowledge

concepts unrelated to value or object in the domain of abstracted scientific knowledge. The next chapter will address the arts and crafts of teaching particular and difficult properties of energy, such as its conservation and degradation. In this chapter the target will be learning to live with abstraction, and to cross over between the two separate domains of knowledge — the life-world and the scientific.

Up in Laputa Land

When Dean Swift lampooned Newton's new scientific philosophy in *Gulliver's Travels* he did so by describing it as a world in the clouds, inhabited by bumbling folk engaging in visionary projects which were quite out of touch with the world. It is a taunt which is often repeated. What truth it may contain is connected with the very basis of the nature of science.

It is an impossibly tall order to outline, in a single section of a single chapter, what is believed about the nature of the process used to construct official scientific knowledge. Nevertheless some rudimentary account is essential if the common difficulties that children have are to be understood. Science is often described as a hypothetico-deductive system. Crudely put, a guess has been made about how things might be at a level where experiment cannot probe directly. This may involve the invisible and intangible; and yet it is thought to explain what is visible and measurable. Gas molecules behave like hard colliding missiles, or electric current behaves like a fluid. From this model for how things are, deductions about phenomena concerning actual gases or real circuits are made. Only then does recourse to experiment for the testing of predictions, add flesh and reality to the descriptive model.

There is no way that these theories and explanations of science, or the concepts of which they speak, can be made to 'drop out' of naïve experimental results. They are hypotheses made by someone who has stood back from the perplexing accumulation of results. The hypothesis is an abstraction, an artefact of imagination, and yet it is used to explain the workings of real things. Logical deduction connects it back to the world of actual phenomena and observations. The philosopher C.G. Hempel wrote of scientific theory that 'it might be likened to a complex spatial network.... The whole system floats, as it were, above the plane of observation' (1952). This bears a distinct resemblance to Swift's absurd Laputa land! Both the separation of the world of theories from that of real things, and the act of 'imagination at a distance' are taxing for children.

Learning to Lose the Context

When pupils begin their first substantial course on energy in the fourth year of secondary school, most will already have begun to move away from an

Getting to Know about Energy in School and Society

exclusively personal view of energy by means of the kind of practical work described in the last chapter. They may also have some idea of energy transfer.

The most demanding task that the physics pupils now face is leaving go, quite deliberately, of the context in which the energy phenomenon takes place. Almost every physics textbook on thermodynamics begins by speaking of a 'system'.

> ...the study of any special branch of physics starts with a separation of a restricted region of space or a finite portion of matter from its surroundings. The portion set aside (in the imagination) and on which the attention is focused is called the *system*, and everything outside the system which has a direct bearing on its behaviour is known as the *surroundings*. (Zemansky, 1957:1)

The 'system' may be an old woman falling painfully down stairs, or it may be a gently flowing mountain stream. For energy purposes, these two situations may become quite indistinguishable. It is not so much, as the text above states, that the attention is focused upon the system, but that the attention is focused only on the energy within the system: everything else is ignored. In this process no detail, no human or other context, is supposed to divert the attention. (The converse will be explored in chapter 8)

There is an opportunity to begin teaching the use of abstracted concepts right at the beginning of the energy course.

WHENEVER WORK IS DONE, ENERGY IS TRANSFERRED FROM ONE SYSTEM TO ANOTHER.

This could be illustrated symbolically

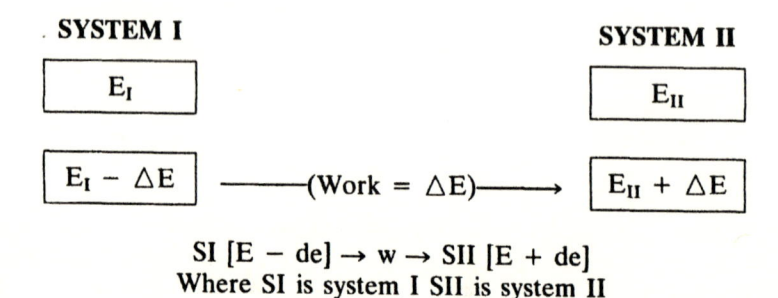

SYSTEM I　　　　　　　　　　　　**SYSTEM II**

| E_I |　　　　　　　| E_{II} |

| $E_I - \Delta E$ | ——(Work = ΔE)——→ | $E_{II} + \Delta E$ |

$$SI [E - de] \rightarrow w \rightarrow SII [E + de]$$
Where SI is system I SII is system II

Correct as it may be, no teacher in their right mind would teach it in that form.

For pupils, energy is often defined as 'the capacity to do work' which is not altogether valid. Sometimes a system, such as a large object at room temperature, clearly has a lot of internal energy since it is not at the

The Formal Domain of Scientific Knowledge

energy). Yet there is no way of getting work out of the system if it is at the absolute zero of temperature (and even there it could have other forms of same temperature as the surroundings. (This is discussed in more detail in the next chapter.) On the other hand the wording of this definition has some merit since it immediately suggests the 'passive' stored aspect of energy which, as we saw in chapter 4, presents difficulties. Whether this definition, or the one which speaks of work as the transfer of energy, is used, the immediate task must be to teach the abstract concept of 'work'.

'Is Work Being Done?'

Teaching that work is: *the force acting multiplied by the distance moved in the direction of the force*, sounds easy enough. The teacher gives examples and the pupils usually seem to understand. Yet the definition has already begun to decontextualize the situation and a simple test can show just how hard this has been to learn.

Two sets of Fourth Year classes, one of higher than average ability and one of average ability were taught this definition of work and then asked:

'In each of the following cases decide if work is being done.'
- a) A woman mowing the lawn.
- b) A heavy weight lying on the edge of a high shelf.
- c) A boy going downhill on a sledge.
- d) A clockwork mouse being wound up.
- e) The same mouse then moving across the floor.

In both ability groups there were very high success rates, of about 90 per cent, in a) and d). But in c) and in e) success dropped to 50 per cent and 60 per cent respectively.

There should have been little difficulty in picking out the cases where force and movement were taking place, yet the pupils chose some cases and ignored others. Work accompanied by effort, in a) and d), were chosen, but where the boy was merely lying on a sledge, or the machinery in the mouse was simply unwinding, it did not seem that work was being done. The pupils observed the details of the system, which they were supposed to ignore, and used their everyday feelings which have no place in a decontextualized domain.

After four week's teaching about both work and energy a similar test was set as an aid to revision. Now the pupils were asked five more questions.

'In which of the following is work being done?'
- a) An apple hanging on a tree.
- b) The apple falling to the ground.
- c) A girl picking it up.
- d) The girl holding it.
- e) A boy eating it.

Overall the results in this test were a little better than in the first. Once again in c) and e) where the work is done by a person the success rates were 90 per cent or above. In b) where the work is done by gravity on 'the system' the success rate dropped to 70 per cent [N=130]. However, the two ability sets did not now perform at the same level. In question b) the top ability groups scored an average of 77 per cent where the lower groups only scored 51 per cent. This significant difference is some evidence that 'losing the context' is a demanding task for all but the more able.

Forms of Energy

Energy can now be defined in terms of work. It becomes a derived, second-hand concept for the pupils so there will be need of much illustration and practice. Energy can be measured in work units, and practical laboratory exercises may follow in which these units become familiar. (These will almost certainly be related to everyday living, and some might almost be about household bills and energy budgets — although these are rarely popular with either pupils or adults, for fairly predictable reasons!) The teacher explains how it is that the operation of work transfers energy from one 'system' to another — from a player to a football, or from a battery to a lamp — then names for the class some of the different forms of energy — kinetic, potential, chemical, electrical, nuclear, etc.

One of the intriguing new ideas for the pupils is that energy can so change its manifestation. It is unpopular in some educational quarters to talk about the transformation of energy from one 'form' to another because it means that energy is not sufficiently abstracted while it is still referred to with these descriptive adjectives. However, the pedagogic value of such words cannot be doubted. If the decontextualization of work already presents problems for some pupils, the complete abstraction of energy, which has so circuitous a definition, is bound to be harder still. Naming the forms of energy serves, at the outset, to help pupils identify the energy being transferred between systems. How else can the two examples above be explained? The energy which has been a part of the system 'player' is clearly of a different genre from the violent movement in the other system — 'ball', or the paste of different chemicals inside a battery to the luminosity of the torch bulb.

Abstract Explanations

Scientific theories are also removed from the events that they attempt to explain. This can be illustrated by comparing the sort of questions that science asks with the child's questions and acceptable answers. Suppose a child warms up a flask of air and watches a tube from it make bubbles in a beaker of water. Continuous play with the apparatus may generate an understanding that it is the act of warming which makes the air expand. This may be a splendid piece of discovery, a linking together of cause and effect into a limited generalization. This is the sort of happy result which makes people exclaim that 'We are all born scientists!'. It may be the starting point for science but it is not yet an explanation of the special kind which science makes.

Why does air expand when it is heated?'. Such a question is much harder, and few children would ask it at all unless prompted by their teacher. Indeed, learning science soon turns into learning to ask questions of this new kind. They are usually answered in terms of a predictive model, and this may be a metaphor drawn from quite another phenomena.

> 'Well, look,' the teacher says, — and it must seem an odd way to answer such a direct question — 'imagine that the air is made up of lots of tiny hard spheres which occupy space because of their rushing movement, and that the collisions they make with the walls of the container which confines them.....'.

The philosopher Mary Hesse has dubbed scientific theory 'metaphorical redescription' (1966). She points out that the explanations which such theories offer do not lie within the domain of the phenomena to be explained but in the very different domain of the metaphor. Sometimes this analogy can get so closely associated with the scientific phenomenon that we may temporarily forget that it is metaphorical at all. Each of the following statements, for example, are basically imported analogies even though they have now become absorbed into scientific thinking:

Sound is propagated by wave-motion.
Gases are collections of randomly moving molecules.
Particles of solute pass through the interstices of the cell wall.

Such statements are very close to being scientific theories. They set the scene, as it were, for the verbal or mathematical argument which constitutes fully-fledged theory. In this final stage of theorizing, which may seem to require no more than the application of logical tools, the model drawn by analogy with everyday things brings with it valuable understanding from the general stock of knowledge.

> The fact that the model system is actually realized in its original sphere implies that its defining properties are coherent and mutually consistent. When, for example, Niels Bohr set up a mathematical theory of nuclear fission, using as his model the break up of a drop of liquid, he did not need to explore his equations in detail to prove that their solutions were unique and mathematically stable; these properties could be taken for granted from his familiarity with the physics of real liquids. (Ziman, 1978)

Getting to Know about Energy in School and Society

The Difficulty with Analogies

If scientific theories are like analogies for real phenomena, why do children, whose own imaginative writing and painting can be so vivid, find them difficult? In what way do uninstructed children explain happenings? Can they handle metaphor or analogy in non-scientific contexts?

Secondary school children are often disconcerting to their science teachers in the way that they attempt to explain scientific events by simple reassertion couched in slightly different words. Ask why sterilization keeps food fresh and many will answer that this 'stops it from going bad'. Or ask why exercise is good for your body and they will write 'because it keeps you healthy'. In everyday speech such reiterations in other words are very common, and there is little doubt that this mode of 'explaining' is taught, albeit inadvertently, to young children by the everyday speech of their own parents.

Many of the questions that very young children ask are not so much requests for reasons and information as calls for comfort and reassurance.

'Why does my tummy hurt?'
'Because you have a stomach upset'
'Why have all the lights gone out?'
'Because we are having a blackout.'

Simple reaffirmation, asserting no more than that things are as they are, is reassuring because it implies familiarity. The very tone of voice suggests that this has all happened before; it was expected — we have a word for it. These examples are not wasted on children. In science lessons and in everyday conversation they continue to use the same kind of non-explanation (Solomon, 1987).

When Jean Piaget investigated the kinds of reasons young children gave for their observations, he listed animism, finalism, and also the kind of reaffirmation that we have just discussed, which he called 'realism'.

> Thus the child invokes as causes sometimes motives or intentions, and sometimes pseudo-logical reasons which are of the nature of a sort of ethical necessity hanging over everything 'it always must be so'. (Piaget, 1926)

The other two kinds of early attempt at explanation which Piaget mentioned — motives or intentions — are also to be found in science pupils' answers. A gas exerts more pressure if the volume is reduced 'because the gas is trying to get out', or, the liquid rises into the syringe 'in order to fill the vacuum'.

All three types of explanation are unsatisfactory in the scientific context because they have a kind of semantic closeness to the event which is

The Formal Domain of Scientific Knowledge

self-defeating. This lack of 'standing back' from the happening means that they can have no interpretative or predictive power in further and wider usage. That is the fundamental purpose of all scientific abstraction.

What useful metaphorical scientific explanations do, is to import an analogical model from some other system in order to get a perspective on the phenomenon. Some writers (e.g. Rumelhart and Norman, 1981) have suggested that most adults and adolescents have already acquired a vocabulary of mental models, arising out of previous physical and social influences, to which they are unlikely to add anything completely new. The process of learning explanations can then only build on existing models by the conscious process of analogy, and then changing them, either by 'fine tuning' or the more radical 'restructuring', as circumstances demand. Most practising teachers take a great deal of this for granted and preface much of their teaching by 'It is rather like a...'.

Children's Use of Simile and Analogy

This approach to explanation hinges on the recipients' understanding of analogy. This is often taken for granted, but an examination of young children's writing shows it to be more complex. Although it is rich in description, often of a comparative nature, exactly the same absence of 'distance' between the object and descriptor can be found there as in the children's science explanations. It is easy enough to write that 'the moon is like a banana', or 'her hair looked like spaghetti'. One visual image just conjures up another which is similar. Gardner *et al.* (1979) have shown that children under the age of 11 who can use this kind of simile are often defeated by more distanced ones. 'The prison guard was like a flint' or 'Electricity is like a river' are types of simile in which the reader is left to find the connecting link — hardness, or flowing — between the two very different systems. Research shows that young children fall back on explaining such similes by close comparison — 'he looked like a flint', or mere juxtaposition — 'electricity cannot be in a river'. It is a real mental feat to move away from the immediate situation in order to track down the appropriate common feature. Indeed, it is regularly used in intelligence tests of the kind: 'Tailor is to cloth, as farmer is to...(plough, sheep, field)'.

Recent work in electricity education has confirmed the existence of these difficulties (Solomon *et al.*, 1987). Classes of First and Third Year pupils were given five similes for electricity: 'Electricity is like fire', 'Electricity is like a river', etc., and asked to give their reasons why they either agreed or disagreed. The data for agreement was interesting, but the lack of ability in handling comparison was far more surprising. An astonishing 58 per cent of the First Year pupils and 39 per cent of the Third Year made at least one glaring mistake in their use of simile.

Nor is it just children who find the active use of simile, in the form of analogy, either difficult or unfamiliar. American college students beginning

105

to learn about electric circuits (Tenney and Gentner, 1984) were first taught about fluid flow in pipes and then about current in simple electrical circuits. Other groups were taught in the reverse order. It might have been expected that work on water circuits would have proved a useful analogy for students' subsequent learning about electrical circuits. No such advantage was to be found. Only if the students were specifically instructed to use the flow of water as an analogy for current were they able to profit from their previous learning. Analogical reasoning does not, it appears, come naturally to the uninstructed.

Where analogical models and simple rules of thumb are both available for learning, their effects can be compared. Teaching that 'doubling the resistance means halving the current' either directly, or by guided discovery through experimental investigation is straightforward enough (Black and Solomon, 1986). The phrase has a good ring to it and proved easy to memorize. Two other similar classes were taught to model 'flow of current' as though it was a fluid, or a movement of electron 'blobs' through a hollow pipe in the same teaching programme. All three groups performed equally well until faced with a new kind of circuit with alternative paths for the current flow. Now it was the pupils with an analogical model, water or electrons, who were most successful.

In the opaque knowledge systems of the life-world, rules without theory are accepted guides to action or non-action. They have, unfortunately, no role in formal science learning since they are context-bound. Only abstracted analogies will have sufficiently wide a currency to be used to make predictions in new cases.

The Theory of Energy Transfer

The pupils will now meet, or construct, some analogy with which to model, in their minds, the transfer of energy. This may itself be taxing for some pupils, and they will certainly have to 'tune' this analogy so that it approximates ever more closely to the accepted properties of energy, such as conservation and degradation.

At the beginning energy is probably thought of as something which can be transferred, passed on, carried by, or transformed. These are the words that the teachers have used, indeed it is hard to know how else they could begin to speak about energy; but as they do so, similes to 'parcels' or 'magical' changes may begin to form in the children's minds. Some physics educators (e.g. Schmidt, 1982) have recommended making this explicit by working through a whole chain of 'energy carriers'. To others (e.g. Warren, 1986) the idea is anathema. It seems as if almost every piece of educational research into children's learning of energy laments the fact that pupils speak of it as though it is 'stuff'. It is not easy to see how else a pupil can begin to accommodate their thinking to an entity (not a word that children use) which is measurable and can be passed on in undiminished quantity. For most of us substance is the paradigm for conservation.

Even if teachers were only speak, as the strictly orthodox would have it, of energy being transferred; still, while they did so, the pupils' thinking would stir. Most people are unable even to hear a nonsense syllable being repeated without forming some mental icon to match it. When words like 'energy flow' or 'transfer' are used we understand them in ways which elaborate familiar mechanisms. Insidiously and inevitably innocent words evoke embryonic models which later teaching and reflection nudge into more acceptable shapes and functions. Adult physicists may possibly end up with the bare mathematical formulation, although it seems somewhat doubtful, but pupils' learning of such an important and difficult concept as energy must begin somewhere. Getting to understand energy is a long and gradual task, during which the pupils' minds play with the idea in progressively more sophisticated ways.

The commonest way of teaching about energy transfer is through diagrams. These are usually drawn, using a system of arrows, to show how this equivalent of work moves about, changing its name and its form as it goes. So the pupils begin to form private images of energy by analogy with some more familiar personal phenomena, aided and abetted by the words of the teacher.

The problem for the pupils is not that the new explanation in terms of energy is difficult, but that it seems unnecessary. Torch bulbs light up when switched on, internal combustion engines run cars to high speeds, rivers in spate flow torrentially down hills. For each of these events there is already a well-known life-world *cause* — an electric current from a battery, an explosion of burning petrol vapour pushing a piston, or the force of gravity acting on water. In most non-scientific situations, to *explain* an event means to find its immediate cause. Only in science (and not, interestingly, in technology) does 'to explain' mean to fit the event into a metaphorical scenario.

Why do They do so Poorly in Tests?

The educational literature is full of complaints or accounts of how little students seem to understand the official energy concept (e.g. Warren, 1986; Watts, 1983). There are also accounts of new teaching strategies which seem to be going well but, unaccountably, have little success in terms of 'correct' ideas delivered in post-course tests. (e.g. Duit, 1984; Brook and Driver, 1986:78; van der Valk *et al.*, 1988).

In chapter 2 this problem of the persistence of children's notions was discussed. 'Conceptual change' is still the unproblematic title of many educational papers and it does seem curious that such a history of failure has not prompted an effort to construct some more viable explanation for what is happening. Where linguistic effects continually reinforce life-world meanings, as is undoubtedly the case with energy, the situation will no longer be one of *do they know or don't they?*, or *has conceptual change taken place?* Old ideas simply cannot die out and be replaced if they are perpetuated through daily talk; and indeed it would be a poor return for a short school course on energy if pupils found themselves unable to understand its life-world meanings. If our pupils are to continue talking sensibly about energy with friends and family they will have two quite different sets of ideas about energy co-existing in their minds.

If this is the case, and there will be evidence to support it, the educational response to incorrect answers to energy questions should be an exploration of why the pupils chose the inappropriate set of meanings instead of the right ones. Context, which was examined in some length in chapter 4, will be part of the key to this; so will the address systems of the memory. It is also the case that there are not just two different meanings but two quite different domains of knowledge involved — one of these expects its meanings to be general and many valued, the other defines them to be precise and invariant. Along with the second are also the beginnings of a new mental model for energy formed by the teacher's words and analogy with other remembered processes.

How Pupils Remember

All teaching and learning rests upon memory and recall, but in the case of energy the process is more complicated. Now it is the retrieval of the 'right' meaning which is the problem. Perhaps it would not be surprising if the two meanings got muddled up, but if they have been learnt and used in quite different situations, as indeed they have, and they belong to different ways of thinking, there is evidence from studies of memory that they might well be separately stored, and be triggered for recall by different cues. Tulving (in Brown, 1960) and others have shown that the context of learning affects the ability to retrieve far more than it does the power of recall. We may have the meaning in our memory but lack the cue which summons it to mind. It is as though the address systems of the two different kinds of meaning are different.

It would be natural to expect better remembering of the scientific concept of energy in the classroom where it was learnt, than in the home where the various life-world meanings are used. The following data (all far more messy and ambiguous than the clean psychological experiments about divers learning lists of words in different contexts mentioned in chapter 4) were obtained in the classroom with pupils who seemed to have difficulty remembering the school definition of energy. In interview and on paper the uninstructed pupils often answered that 'work' meant the job you got, and 'energy' how fit you were. The new physics concepts were carefully talked through and tested until well over 70 per cent of the pupils could respond correctly when cued by the special environment of the physics laboratory.

Within a few months it was sad but not very surprising to find that many

The Formal Domain of Scientific Knowledge

Table 3   Cross-tabulation of pupils' responses to last questions. The figures give the numbers of answers received

		Responses to 'Examples of Energy' question				
		Life-world	Mixed	Physics	Omissions	Total
Responses to 'Energy Changes' question	Life-world	0	0	1	0	1
	Mixed	0	0	1	0	1
	Physics	20	4	59	5	88
	Omissions	10	3	21	6	40
	Total	30	7	82	11	130

Source: From 'Prompts, cues and discrimination: The Utilization of Two Separate Knowledge Systems', *European Journal of Science Education*, 1984, Vol. 6(3), p. 279.

of the less able pupils appeared to have forgotten the new physics meanings. Once again 'work' just meant jobs, and 'energy' just meant being energetic. Presumably frequent use of these words in the home context had continually recalled and reinforced only the life-world meanings for them. Little wonder then that the meanings learnt a couple of months before in the physics laboratory were hard to retrieve. This is a depressingly common event in school tests.

Was it a problem of retrieval or recall? If the pupils had been storing these physics definitions separately from the life-world ones, they might simply need a different cue. Two different questions about energy were now put to the pupils. In the first they had to give three different 'examples of energy'. In the second they had to name the 'energy changes' that took place. This second question added a recognition cue through the use of physics terminology — '*energy* changes' (Table 3).

1   Only 3 per cent of the total responses contained mixed answers with life-world examples (e.g. a boy running, or water behind a dam) alongside the physics terms for energy (chemical energy, kinetic energy, etc.)
2   69 per cent of those who attempted both questions gave physics answers to the 'Examples of energy' question, while an impressive 98 per cent of them used physics terms in response to the 'Energy changes' question.
3   Out of the 30 pupils who gave only life-world examples for 'Examples of energy', 20 gave the physics terms when triggered by the phrase 'energy changes'.

These data begin to sketch a picture of separately stored stocks of knowledge which can be brought to the surface by different triggers. Apparent forgetting may thus be no more than the lack of the right recognition key to unlock the memory file.

This impression was reinforced by evidence from individual interviews.

Getting to Know about Energy in School and Society

In the unfamiliar, and inevitably slightly tense, situation of an interview there were often long pauses when the pupil's mind 'went blank' and they could not recall some point. A reference to physics by the interviewer could not only trigger a list of appropriate learnt energy concept names but also showed that these arrived fully clothed with correct exemplars from the everyday world.

Teacher.	If I ask you now for examples of energy, what will you give me?
Pupil.	Um...energy from a battery.
T.	Yes?
P. (Long pause)...	I really cannot think of any now.
T.	Did you learn any in physics? (Pause) Different kinds of energy?
P. (quickly)	Well, I know what potential is and nuclear and kinetic.
T.	What does potential mean?
P.	Potential, that's something high up like an HPS station.
T.	Yes, that's quite right — and kinetic?
P.	Anything moving.
T.	Yes, give me an example of something moving.
P.	Somebody walking along a road and, er, the piston in a car.

The same sort of triggering action which brings into play a different store of information is also common in classroom discussion. Either the form of the opening question by the teacher, or a single answer by one of the pupils, may act as the recognition and retrieval trigger. The following two trails arose, one after the other. The first question was 'What sort of things have energy?'

→(batteries)—(electricity)—(a hammer up in the air)

'Is that all the forms of energy you can think of?'

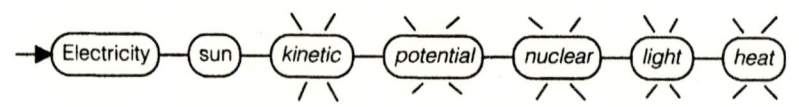

→(Electricity)—(sun)—(kinetic)—(potential)—(nuclear)—(light)—(heat)

The Two Ways of Knowing

Not only are the two kinds of meanings separately stored in the brain — the mere fact that they were learnt in such different circumstances almost

The Formal Domain of Scientific Knowledge

ensured that this will be so — they also involve radically different kinds of thinking, the scientific and the life-world.

The best way to illustrate what this might mean is to set a task which can be tackled in two different ways corresponding to the two different knowledge systems. The first question is given below.

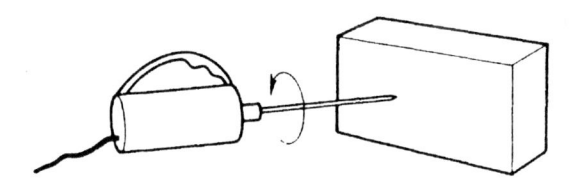

'An electric drill, working at a rate of 500 watts, is used to drill a hole in a piece of wood. How much work could it do in 20 minutes? *What energy changes are taking place?*'

It is the last part of this question which demands, at the very least, one movement of mental operation from the life-world which is called to mind by the picture, to the scientific.

In the event it turned out that there were three general ways in which pupils tackled this part:

1. By giving the life-world causes of each part of the drill's working, e.g. the electricity makes the drill rotate which makes a hole in the wood and makes heat by friction.
2. By giving the formal sequence of scientific redescriptions of the energy, e.g. electrical energy — kinetic energy — heating energy.
3. By giving *both kinds* of explanation alternately and relating them to each other, e.g. 'Electrical energy goes into the drill and is turned into kinetic energy as the drill goes round, then friction between the drill and the wood turns this energy into heat energy.'

The first kind of answer shows no movement at all out of the life-world domain. The second requires one translation of thought from the life-world domain of the question as it is set, to the scientific domain of energy transformations. However, any pupil who can answer in the third way shows an impressive ability to move in both directions between the two domains, correctly and at will.

Crossing between the Two Domains

What sort of capability were these pupils demonstrating by this display of mental flexibility? There was no question that those who answered in the second way were correct. Was the third way of answering, which was only attempted by a mere 11 pupils out of 42 in the top ability classes, actually better in some sense? It was easy enough to show that this special group of pupils did not score significantly higher than the others in the test from which this question was taken.

However, when 2½ months had passed and the same kind of question was set again, another difference in their learning came to light. As usual a depressing number of the pupils had either forgotten, or had allowed life-world knowledge to smother the physics that they had learnt. The number of those answering in exclusively life-world terms rose from 19 per cent to 36 per cent. What was significant and interesting was that only one pupil, out of the sub-group who seemed to have mastered the art of movement between the domains, reverted to the life-world way of thought. Successful crossing over and back from one domain to another may indicate that more durable learning has taken place, and, perhaps, that a deeper level of understanding has been achieved.

In the following year, when all the pupils were taught to answer this kind of question in both ways, deliberately crossing between the two domains, their overall level of success in the two tests increased dramatically, from 57 per cent and 86 per cent, to 78 per cent and 96 per cent. A somewhat similar effect was achieved by Brook and Wells (1988) who wisely encouraged their pupils to discuss each energy change in groups thus ensuring that pupils used both domains of knowledge. In this way social construction could aid in the difficult crossing from the life-world to the science world of meaning and back again.

An Asymetrical Effect

Crossing between domains sounds as if it is no more than a question of mental flexibility, but some test results suggest that not all crossing operations are equally difficult. Whereas life-world knowledge is 'learnt' through social reaffirmation in everyday situations, the more esoteric knowledge of science is the product of school learning — a later, secondary process of socialization. This makes it more fragile than the robust social structures of common sense and less imperative in seeking attention.

> The life-worldly stock of knowledge is confirmed in the mastery of situations,...Otherwise put, the 'suspicion' of a fundamental inadequacy in the life-worldly stock of knowledge cannot arise in the natural attitude. (Schutz and Luckmann, 1973:171)

What is required in school learning is a very special attitude which can allow what Schutz has called a 'leap into another province of reality' — the new

The Formal Domain of Scientific Knowledge

domain of scientific thinking. Any subsequent lapse of attention in this special scientific attitude may trigger an easier leap (perhaps it might be better described as a 'fall') back into the intuitive thinking processes of the life-world. Thus we shall expect to see, in the work of our pupils, special difficulty in reaching the symbolic domain of science, some skill in performing work while there, and a much easier return journey back into the arms of the life-world.

These expectations are not only borne out and illustrated in the school learning of physics; they also serve to explain results obtained in several other independent pieces of research.

1. The pupils tried two questions:
   In the first they were given the picture of an electric appliance and asked to name the energy changes going on. (L-w to Science)
   In the second they were given the energy changes and asked to name the device to which they referred. (Science to L-w)
   In every one of four classes the second question was far better answered than the first (60 per cent to 90 per cent success)
2. In Jill Larkin's work on the problem solving of novices and experts (1980) it was the initial move of the expert into the domain of abstract thought, not the subsequent interpretations of theory, which most clearly differentiated their procedures from those of the novices.
3. Lawrence Viennot's study of dynamics (1979) showed that physics graduates did much worse when a problem was specified in the life-world (a juggler throwing balls into the air) than when couched in the abstracted language of the physics domain. (e.g. 'If two equal masses are acted upon by identical forces...'). No domain-crossing was needed.
4. Pupils were given a series of 'right' and 'wrong' sentences to sort out. There were questions about units, definitions, and simple calculations. Just two contained inappropriate movements from the domain of physics into that of the life-world.

   'The kinetic energy of a car turns into friction when it brakes,' and
   'The water at the top of a waterfall has potential energy which turns into spray at the bottom.'

In both of these, the sense of the statement falls back inappropriately into the life-world in the second half of the sentence. Only a minority of pupils in the top ability group, and none at all in the lower group, could detect that any transition had taken place at all. By subdividing the pupils by ability (examination score) it was possible to show that these two questions were better performance indicators than any of the others, even those which required mathematical calculation.

Living in Two Domains

It would be quite wrong to suggest that science is the only structured abstracted system. Nor is it the only one which is at odds with another more anarchic life-world domain. Verbal and written text, moral maxims and ethics, religious practice and theology, are other pairs which suggest themselves. It may be worth using the work of an eminent anthropologist to indicate that the tension between the twin but opposing urges to live by accepted rules of thumb, and to formulate systems of logical and codified rules to curb and restrain living, actually reproduces a two domain effect in legal terms. This is how Pierre Bourdieu discusses the domains of custom and law in a North African community.

> ...customary law always seems to pass from particular case to particular case, from specific misdeed to the specific sanction, never explicitly formulating the fundamental principles which 'rational' law spells out explicitly (e.g. All men are equal in honour)...Thus the precepts of custom, very close in this respect to sayings and proverbs, have nothing in common with the transcendental rules of a juridical code. (Bourdieu, *Practice and discourse about practice*, p. 17)

The community Bourdieu was studying did possess such a transcendental juridical code containing abstract universal statements like 'All men are equal in honour' at the same time as customary law. If that passage spells out, as I believe it does, the anthropological equivalent of two domains of knowledge — the life-world rules of thumb, and the universal scientific principles — then we may be drawn into seeing the two domain effect as an inevitable concomitant of human living and rationality. On the one hand there will always be the gathering together of a stock of day-to-day knowledge; on the other there is a perennial search for higher level order and explanation. Neither, however, precludes or supersedes the other. The intellectual often lives by custom, just as the layman may ponder universal structure.

The question is no longer why science is so different from general knowledge, but whether higher order knowledge of the natural world, which is bound to be context-independent, is the only effective domain for scientific thinking. Is that difficult crossing over between domains, from commonplace reality to distanced analogy and back again, essential to scientific thinking? The philosophy of science suggests it is, and yet reflective teaching

The Formal Domain of Scientific Knowledge

points out its considerable difficulties. Of one thing we may be sure, this movement between the two contrasting ways of thinking requires dedicated practice rather like athletic training. But if our pupils are to be allowed a glimpse of how science constructs its concepts and explanations, then such exercises may be the only way to achieve the necessary mental insight and flexibility.

Chapter 7

Teaching about Conservation

About Teaching

The next stage in the energy course was to bring in ideas about the conservation and degradation of energy. This involved an experiment in teaching where the traditional order was reversed and a completely new teaching progression was introduced.

Is there, or could there be, a reliable model for effective science teaching? The work of Shayer and Adey (1980) categorized concepts by their conceptual difficulty in the Piagetian scheme to show what a teacher would be advized *not* to attempt, but gave no guide on how teaching should take place. Their later work (Shayer, 1990 op. cit.) reported a number of successful outcomes in terms of cognitive acceleration, but still achieved no general model of teaching as a process, rather than learning as an outcome. This problem is sometimes known as 'theory into practice' and has also been analyzed by Rowell and Dawson (1989) from three theoretical perspectives — cognitive psychology, artificial intelligence and Piagetian epistemology. They drew generously upon a whole range of research findings, but whether their integration is equivalent to a teaching theory was left as an open question.

Teaching and research have always lived uneasily together, with the researchers worrying continually why it was that teachers did not welcome their ideas and put them into immediate action, and the teachers maintaining a withdrawn silence. The impasse is probably due to at least four factors:

- professional distrust or blame between teachers and researchers,
- lack of communications (teacher/teacher, and teacher/researcher),
- uncertainty about modelling the teaching-learning connection, and
- the different kinds of knowledge about the learning process that teachers and researchers possess.

The first and second of these seem as though they could be cured by personal management techniques, and indeed concerns about them have prompted collaborative research in several countries, notably Britain with Adelman and Elliot in the University of East Anglia, and New Zealand with Baird and White at Monash University. Both projects have had some successes — certainly more than the 'mastery' attempts of Bloom and Carroll in the United States during the 1960s. But the basic question of defining teaching has not been answered.

Consider the third problem listed above. It is natural for the community of researchers to begin with the learning process and move on to teaching. Indeed, it seems quite the most logical pathway to take. Only on closer examination do doubts arise. The constructivist school of research have clearly shown that children bring their own unorthodox explanations and descriptions of common phenomena to science lessons. As we have seen in chapter 2 the simplest learning approach to be based on constructivism, which encourages the pupils to 'test' their own ideas through their own laboratory explorations, does not often achieve success. Hewson's highly rational criteria for the promotion of learning by means of new ideas which are *plausible, intelligible and fruitful* for the students (Hewson, 1981) ensure no better outcomes. Children may still hang on to old familiar notions.

The most elaborate learning model, which takes into account the early work of Ausubel (1968) on how previous ideas organize new observations, and the Gangé and White (1978) on memory structures and learning outcomes, as well as the now well-established notions of constructivism (*see* chapter 2), is Osborne and Wittrock's *Generative Learning Model* (1985). It was based on learning research, and also on evidence acquired from classroom observation. The picture of learning it produced has a coherence and flexibility which few have matched.

This model (p. 118) is highly successful in explaining *why teaching might not work*. The authors put it in this way:

> The model also explains...the inordinate influence of existing ideas in determining
> i) what sensory input is selected and attended to,
> ii) what links are generated to existing ideas,
> iii) what meanings are constructed,
> iv) what evaluations of new constructions occur, and
> v) what constructions are subsumed into long-term memory.
>
> ...Both no-learning and unanticipated-learning can be accounted for. (69)

The model describes learning, and why teaching may have failed, if it does. It is *not a model of teaching*. Analogies with other fields of practice demonstrate the same kind of problem. Models which describe economic systems, social interactions and human medical functions, can all be used to

Getting to Know about Energy in School and Society

Schematic representation of the generative learning model

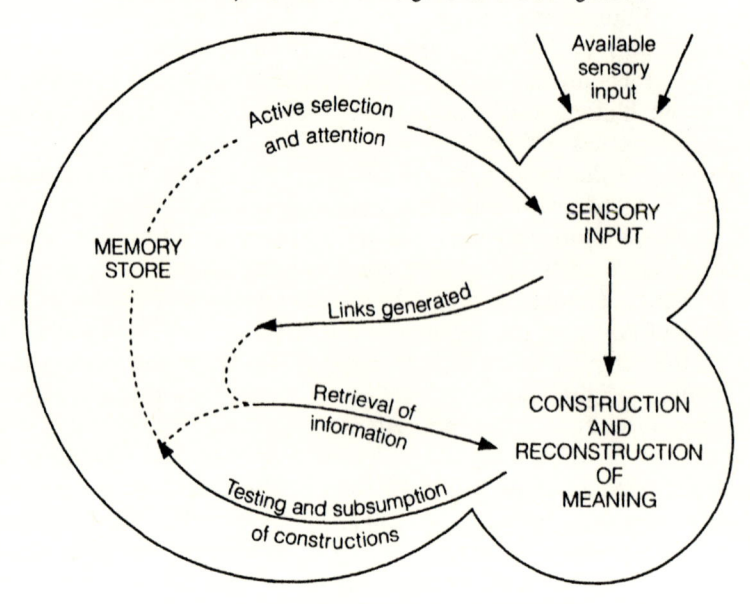

THE ACTIVE CONSTRUCTION OF MEANING

track down faults in the relevant system, just as the generative model does for classroom learning. Unfortunately, they do not show how to bring about economic success (alas!), get to be popular with other people, or construct a living organism. On the other hand a mechanical system, like a car, may be described so completely that not only can faults be tracked down and put right, but it would also enable a new car to be constructed.

Teaching, like the examples given above, is a human process which is almost indescribable in its complexity. Only its outcomes can be described and even explained, in terms of learning research. Perhaps teaching should be thought about as one would a travelling process. The teacher's professional task is to sense the route to take. She may use research findings about learning for planning the lesson but her tacit craft knowledge — the professional reaction to continual feedback from the familiar faces around her, and past experience about what helps these particular pupils to give their full attention to the learning task — will provide such strong influences upon teaching that they defy definitive modelling. The teaching and learning relation certainly cannot be described by causative mapping.

The fourth reason given for the teacher/researcher unease — the different kinds of knowledge used to evaluate classroom evidence — is best left to the end of the chapter when findings will have been displayed.

Scandals of Energy Teaching

Even within academic physics the energy concept is fraught with difficulty. The greatest modern teacher of university physics was undoubtedly Richard Feynman, who simply refused to define the term energy. In his renowned *Lectures on Physics* he wrote:

> ...there is a certain quantity, which we call energy, which does not change in the manifold changes which nature undergoes. That is a most abstract idea because it is a mathematical principle. It is not a description of a mechanism, nor anything concrete: it is just a strange fact that we can calculate some number and when we finish watching nature go through her tricks and calculate the number again, it is the same. (Something like the bishop on a red square, and after a number of moves — details unknown — it is still on some red square. It is a law of this nature.) *Since it is an abstract idea, we shall illustrate the meaning of it by an analogy.* (my italics) (Feynman, 1963)

The analogy that Richard Feynman then used is great fun. It features 'Denis the Menace', his friend Bruce, a toy box, some dirty bathwater, and Denis' mother who patiently devises ever more complex formulae to keep track of the 'energy bricks' which the dreaded Denis keeps trying to conceal or chuck away. In this way Feynman left the direct abstract route to understanding, and used an accessible analogy. His brilliance as a teacher was attested by the knowledge of his students, and also in their enjoyment of physics.

By way of contrast, other educators have seen in the abstraction of energy a reason for teaching it in a severely abstract spirit.

> Ideas must be presented clearly, explicitly and precisely. Any vagueness, ambiguity or contradiction causes endless difficulty.... One can only begin to learn about energy when one understands work. Work is an abstraction from the quantities displacement and force. Force is an extremely difficult abstraction which can only be taught on an axiomatic basis. Sensations resulting from the deformation of our bodies must not be regarded as 'experiences' of forces.... (Warren, 1986)

In the school described in this study, formal teaching about the concept of energy began with the measurement of work, as Warren stipulated, but it included plenty of practical investigations which did involve body experiences like pulling a trolley up a slope. Earlier practical investigations had used other body experiences, like winding up a weight. Neither students nor adults begin their learning easily through abstractions, however 'precise and unambiguous'. The rationale for this is not far to seek. Abstractions themselves are collectives from multiple experiences on a common theme, in much the same way as the mathematical concept of 'two-ness' is refined out of many concrete experiences of two objects — two bananas or two archbishops. Only later does it become possible to see the examples as incumbrances rather than aids. This is the moment at which the ladder to the life-world is kicked away and the abstract concept becomes self-supporting.

The words we use when teaching children to recognize energy situations, like 'chemical energy', 'kinetic (or movement) energy', 'electrical energy' or 'light energy', have also come under unnecessary attack from the physics purists, but none so strongly as 'heat energy'.

> Use of the word *heat* as a noun should be avoided. We should always try to use the word *heating*. (Summers, 1983)

> The term *heat* is perfectly adequate so long as attention is drawn to the difference between a state and a process. (Mak and Young, 1987)

> *The cooker supplies lots of heat energy*...carries an implication of heat as being a separate and distinctive kind of energy, revives notions of 'caloric' and implies that energy is a substance. (Osborne and Freeman, 1989)

In the last case research into children's notions tells us otherwise. As we have seen, pupils do not commonly understand how energy can be stored in a passive form, before they have been taught about it. None, in my experience, speak of it as a solid substance or have even heard of the eighteenth-century term 'caloric'. By contrast the bricks of Denis the Menace are unashamedly substantial, and it may also be noted that Feynman had no qualms at all about using the term 'heat energy' with his American college students.

Methods of teaching are clearly extensions of teacher personality and style; but that does not mean that they are arbitrary. Through listening and questioning, teachers worthy of that name adjust and change, almost continually, what they say and the investigations that they set, in order to match the growing understanding and enjoyment of their students. This is professional craft knowledge, but when the professionals are teachers without the status of a Feynman, the knowledge is thought so little worthy of reliance that almost any educator seems to be able to over-ride its insights.

Transferring Energy

It is a fundamental property of energy that it can move from one system to another, and that it is conserved. At least three difficult problems arise when trying to understand this.

a) Energy moves in certain directions and not, with equal facility, in others.
b) Energy is often dispersed, and things seem to run down.
c) Frequently, and more mysteriously, energy may seem to have been transformed into a different manifestation.

The last point, linked as it is with the second, differentiates the energy concept from many other conserved quantities like mass or momentum. It is for this reason that we name the different types of energy for our apprentice physicists so that they watch out for the energy (a good deal more difficult task than Denis' Mum's effort to keep track of his bricks!). The value of energy descriptors in teaching is inestimable because they help the children to follow, in their conceptual 'mind's eye', the transference of this elusive entity from one system to another. When chemicals ignite and engines move we need a strong rhetorical language which will speak of the light and heat, of the motion and noise, as the vivid concomitants and outcomes of this energy transfer. If we fail to provide this language our pupils will revert to a life-world contemplation of simple 'burning' and 'chugging' (as we found in the last chapter) in which there is scant room for any energy concept at all. More serious still, they will make no progress towards any idea of the continuity of energy which is the first step towards understanding its nature and its conservation.

In the last chapter the pupils began the difficult process of abstraction. In this chapter the purpose of this becomes clearer. The pupils are guided towards answers to the following questions:

1. Does energy increase, diminish, or stay the same? (Conservation)
2. In which sense or direction is energy transferred? (Dissipation)

These can only be answered if we follow the abstracted energy entity from one system to another, and from its first manifestation to a second or a third.

Intuitive Knowledge about Conservation

The first step in understanding conservation is to believe that energy is storable so that it can be thought of as a continuing commodity. Qualities like human 'energeticness' which may be little more than a feeling, do not seem at all likely to fit. Movements of all kinds tend to slow down and stop.

So it came as some surprise to find that when Third Year school pupils, who had so far no instruction in physics, were asked whether they thought that energy could be stored 81 per cent of the sample replied that they thought that it could be.

Closer inspection showed, however, that very little weight could be placed on this raw statistic. Those who gave 'food' or 'fuel' as examples in their earlier statements about energy were, curiously enough, no more likely to say that energy could be stored than were those who gave movement examples. Similarly, there was no significant correlation between making a universal statement about energy (see chapter 3) and believing in the storage of energy. Indeed, reflection suggests that even splendidly decontextualized individual constructions like 'energy is what makes things work' do not imply its continued existence after the movement has finished. Possibly the running down of movement might well suggest that it can never be stored: friction is an inescapable reality of our world. Finally, there was not even a significant negative correlation between having given only human examples of energy and believing that it was storable.

Young children, as we have seen in chapter 5, readily learn about energy storage and transfer through making models of 'wind up' machinery. Nevertheless, many Third Year pupils were puzzled about how to respond to the question — and with good reason. The answer that may have got nearest to expressing the common dilemma about energy conservation was given by one of these uninstructed pupils:

'Energy can be stored but it decays in a sort of way and has to be kept *virulent*. Although some energies can be stored more easily.'

Examples given of the storage of energy ranged from batteries, foods and fuels, to less convincing items like engines and power stations. There are little grounds here for believing that the question had been adequately understood.

The Fourth Year pupils who were encouraged to explore their own ideas about energy at the start of their formal instruction offered similar ideas when they spoke about what the storage of energy meant.

'The safe-keeping of energy.'
'...controlling energy.'
'...not letting it out...little bits at a time.'
'...holding it in.'
'Letting it out slowly.'

These extracts certainly make energy sound difficult to store and perhaps even dangerous. They may be consonant with the notion of energy as a symbol for activity verging on destruction which was met in the adult conversations (chapter 1). They also suggest that storing a quantity with the

Teaching about Conservation

attribute of energeticness is likely to be a difficult, almost impossible, task — like the caging of a wild animal. The subject has been well illustrated with research data in chapter 4.

This question — whether energy continues to exist or dies away — was also answered differently in the history of science according to intellectual inclinations. Galileo's ideas often took off from some interesting practical observation. His work on the strength of materials began by watching boat builders; his work on water pressure from a chat with the workman who was mending his pump. For Galileo the notion that some function of height and motion might be conserved was inspired by watching how a pendulum, which hit against a peg, reached the same vertical height as it had started from, by another route (below).

Now, gentlemen, you will observe with pleasure that the ball swings to the point G in the horizontal, and you would see the same thing happen if the obstacle were placed at some lower point, say at F, about which the ball would describe the arc BI, the rise of the ball always terminating exactly on the line CD. But when the nail is placed so low that the remainder of the thread below it will not reach to the height CD (which would happen if the nail were placed nearer B than to the intersection of AB with the horizontal CD) then the thread leaps over the nail and twists itself about it.

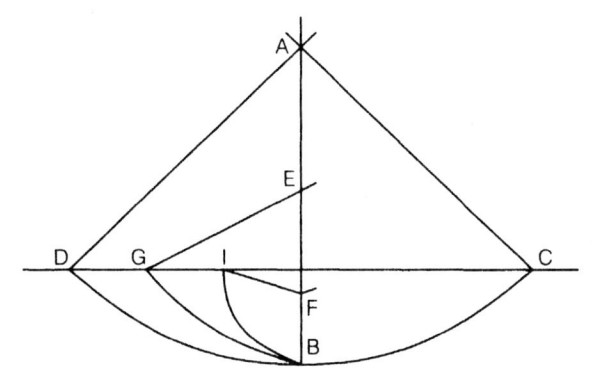

Galileo's pendulum

Source: Galileo, *The Two New Sciences* (Discourses on the Third Day).

Leibnitz, a century later, was an archetypal abstract mathematician. His writings give the impression that the energy of motion needs to be reified and conserved for some purely conceptual reason. So he just ignored the 'running down of things'. This attitude exemplified the sad schism between the abstract theoreticians and practical thinkers which has existed down the ages in western Europe. We may indeed have glimpsed it again in the present day controversy over teaching about energy.

Conservation and Reversibility

Understanding the conservation of simple concepts has always been a high profile educational achievement. It figured prominently in Piagetian tests of cognitive development when applied to simpler concepts such as volume, number and mass. During the last ten years there has been a continuing academic controversy about these tests and the language used to administer them. What do pupils understand when the questioner asks if there is 'the *same amount* of water' in a tall and narrow container as in a wider shallow one? There is a real sense in which they are not the same; one is clearly deeper than the other. Unless the child already has an idea about what volume means, the question makes little sense. It would be a worrying situation if only after the concept is understood can questions about it be understood and answered. On reflection, this situation is hardly surprising to those who understand the role of language as an exploratory tool. In more recent 'child-centred' investigations enormous quantities of orange juice have been poured to and fro, by tester or tested, to answer the question 'how much is there now to drink?'.

The point of interest for us in these Piagetian tasks is that there is a half-way house to true understanding of conservation in which the child states that there is less (or more) of the fluid now, but that the situation is reversible. If it were poured back into the original container, they concede, its volume would then be the same as it was at first. Some psychologists see this as no more than the child's use of words like 'more or less' to describe the superficial appearances of height and depth. They believe that the child is already conserving the liquid's volume, but is describing its level. Others, along with Piaget, maintain that since reversibility is not the same as conservation no such conclusion about conceptual understanding can be drawn.

Whichever side of this controversy wins the argument it is clear that these 'reversing' children have made significant progress in mental modelling. They can operate mentally, manipulating the quantity of orange juice in their minds, and hence make forecasts about what might happen if it were poured back. Such mental operations with a concept or theoretical model are central to physics. They figure in several branches of modern cognitive science as well as owning a long ancestry in the *gedanken* or thought experiments which have enabled gifted thinkers, like Einstein or Archimedes, to make their great theoretical advances. Above all they are used to make predictions. In this case they predict a reversible situation.

Unfortunately reversibility is *not* a property of energy transfer. Worse still, the conservation principle, which all educators see as central to the nature of physics, is poor in observational predictive power because it so readily changes its manifestations. For these reasons the second question posed about energy — the direction of its flow — is one with great significance and practical importance. Petrol burns, heat flows, and buildings topple. All such processes release large quantities of available energy but are *not* reversible. Cognitive progress towards understanding conservation, via the common staging post of reversibility, is clearly not open to pupils in the field of energy..

The following extract, which illustrates this point, was taken from a classroom discussion by a group of middle ability pupils who had just begun learning the Principle of Conservation of Energy. The ludicrous notion of energy release during combustion being reversible is justly quashed by the last pupil's comment. The really extraordinary fact is that this was by no means a solitary example of belief in reversibility. The pupils' written answers to a question about conservation to be reported later, show that it is a fairly common way of coping with a notion of conservation which is not fully understood.

Teacher.	*What happens to the energy in the fuel we use?*
P1.	It's changing, changing.
P2.	Changing into moving energy.
P3.	It hasn't been replaced.
T.	*Can we actually run out of energy?*
P1.	No you can't, can't destroy it.
T.	*Can we run out of fuel?*
Ps	Yes...yes...yes....
P3.	Then you change the energy back.*
T.	*Can you change energy back into the fuel then?*
Ps	No...no....
P3.	Yes you can.*
P1.	If you could, you'd be a millionaire!

What to Teach First — Conservation or Dissipation?

The story of how energy came to be recognized as a conserved quantity spans several centuries and is described in more detail in chapter 9. The mathematical equivalence of work and heat, and the conservation of the new combined quantity, became established before any theory of working heat engines was accepted. Hence the First Law of Thermodynamics, describing the conservation of energy, arrived first. The Second Law, describing the dissipation of energy, followed some way behind the development of the steam engine. The dissipation, or running-down of energy changes, is not confined to engines but it was their inescapable presence in factories and trains which stimulated the first mathematical interest in this concept.

Historical order, however, is no guide to how modern children might best learn about energy. The 'primitive science' of great thinkers of the past was far too cogent and philosophical to match children's naïve explanations of how things might be, and the commonplace objects of our age are also very different from those of earlier times. We would not expect ancient imaginative models, like 'internal tremors' or 'weightless fluids', as explanations of heat from our modern children. We could also question whether the First Law, about energy conservation, precedes notions of the Second, about dissipation, in our pupils' thought (Solomon, 1982). They have grown up with cars, engines and motors. It would be surprising if their intuitive thermodynamics did not try to explain and predict the course of energy transfer and transformation in some simple way.

Trying to Learn the Conservation Principle

Teaching a student to understand and apply any physics principle involves two processes. The first of these is finding appropriate language for the task. This, in turn, requires both that the child will understand the form of words used, and also that these words describe an aspect of the principle which the child is able to grasp. It is a process of bringing the child and a simple aspect of the principle, closer together.

The second essential part of planning the teaching process is taking account of the common intuitive problems of the children and arranging the illustrative examples, and the order of teaching, so as to be as satisfying and efficient as possible. It is not so much a process of displaying the logical progression of our own thought — a pitfall of egocentricity which may trap the very best of physics teachers — as of arranging a rhetoric which the child will find convincing.

Neither of these precepts is often followed in the conventional mode of teaching the Conservation Principle. Fourth Year pupils in the pilot study which preceded the main research work into the learning of energy, were simply told the Principle, as usual, followed by a brief explanation. They then had to copy down the form of words in the time-honoured way, and learn it.

> *Energy is never created or destroyed but can be changed from one form into another.*

There is something in the cadences of the Conservation Principle phrased in this way, which makes it very easy to memorize. Unfortunately this does not imply that it will be understood. To probe what this form of words meant two questions were set. The pupils were presented with them in each of two consecutive tests.

A. Explain what is meant by 'The Conservation of Energy'....

Teaching about Conservation

The car is travelling along when, suddenly, the driver finds he has run out of petrol.
a) Has the car got any energy left? ....
b) If so what will happen to this energy? ....

B. State the Principle of Conservation of Energy.

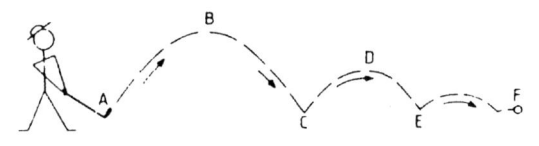

Sketch of golfer and bouncing ball

The golfer in the sketch has just hit the ball along the path shown so that it stops at F.
Describe *all* the energy changes that have taken place as completely as you can ....

The pupils' responses were then used to answer the following questions:

- Could they recall the Conservation Principle?
- How did they answer the part about energy changes?
- Was there any significant association between recalling the Principle and giving a correct answer about energy changes?

There were six different ways of answering A, part b), and also question B:

1. Omitting it altogether.
2. Giving a life-world answer such as 'the energy is lost', or 'it slows down and stops'.
3. Assuming the Principle to mean that the energy is stored somewhere — in the car's batteries or in the ball.
4. Assuming that conservation implies reversibility — 'it turns back into kinetic energy'.
5. Repeating the last part of the Principle, it is 'changed into another form'.
6. Using the Principle to explain the stopping of the car by a transformation of its kinetic energy into another form of energy, such as heat in the brakes or in the ground, which then leaks away.

As might be expected, most of those whose answers were in categories 1 and 2 had not recalled the Conservation Principle. Without its help the chances of getting the right answer were sadly small. But the results of the other pupils, the clear majority of the class who had learnt the Principle and

127

Getting to Know about Energy in School and Society

Table 4 Car running out of petrol problem

Class	Omitted	Life-world[††]	Storage or reversible	Transformation	Transformation + dissipation
Top ability (25)	6	4	7	5	3
Middle ability (21)	6	6	4	3	2
Total	12	10	11	8	5
Conservation Principle correct	7	3	10	7	5

[††] These answers merely stated that the energy was gone, lost or finished.

Table 5 Golfer's bouncing ball problem

Class	Omitted	Life-world[††]	Storage or reversible	Transformation	Transformation + dissipation
Top ability (24)	0	2	10	6	7
Middle ability (22)	8	2	4	8	0
Total	8	4	14	14	7
Conservation Principle correct	0	2	12	10	7

could recall it correctly, were also often wrong, albeit in another way. Out of the pupils who answered the questions in categories 3, 4 or 5, almost all (85 per cent and 83 per cent respectively) gave the right form of words for the Conservation Principle. It seemed that knowing this did very little for their chances of success. It was even possible that recalling its form of words (as in the case of reversibility already encountered) was actually giving pupils the wrong clues.

Only by listening to the words of the pupils can the teacher distinguish between these two possible impediments to understanding. In this example three girls were examining their marked work and talking together.

> T. *Did you really mean that all the energy was stored up in the golfer's ball?*
> P1. Yes, I did. I mean that's conservation, ain't it?
> T. *But it makes it seem as if the ball is a bomb, with all that energy stored up in it! Did you think that?*
> P2. Yea, I thought so too. It is 'not created or destroyed', so it must be there still.
> T. *In the ball, Angela?*
> P3. (Uncertain) Yea, I thought so.

So it began to seem as if the first requirement for good teaching — that the pupils should understand the words used in the sense that was intended — had not been satisfied.

More evidence for this was the lack of any significant correlation between knowing the Conservation Principle and answering the question correctly (answer 6 above). Other questions in this mechanics test had been set in the same format: they asked for the recall of a concept or principle first and then set a problem in which it might be used. One of these was on stable equilibrium where the problem required a drawn solution; the other was on acceleration and needed a simple calculation. In both cases the correlation between successful recall and successful problem-solving was very high (better than 0.80). Only in the energy question was there no significant correlation between recall of taught principle and problem-solving.

One explanation for failure was that these pupils had learnt the Conservation Principle rote fashion and completely failed to understand what it meant. Another possibility was that the Law was taken to imply reversibility. A third reason could have been the poor predictive power of a Conservation Principle which says nothing about the new form or location of the energy.

A Change in Teaching?

Teaching is an absorbing but developing profession. Student teachers are rightly expected to plan out their lessons on paper and then to carry out the plan unchanged while their supervisor or mentor watches, but most mature teachers behave in an outrageously different fashion, making changes as they go along. They use the crafts and tools they know — methods and 'teaching wrinkles' — much as a carpenter might use a chisel, while recognizing the problem of knots in the wood. Teachers have both well-trodden teaching paths, and also the irritant of well-known pupil pitfalls. The commonest resolution of these tensions is to teach as before, but add a warning to the pupils about not 'getting it wrong'. It is the wash of classroom feedback which stimulates this simplest kind of 'reflection-in-action' of which Schon (1983) wrote. It mitigates the teaching programme, guiding the chisel in a slightly different shaping movement; but because it takes place mid-lesson it can never set out to re-shape the whole teaching approach.

Designing a completely new programme requires a strategy thought out before the course begins, rather than small remedial tactics in response to pupil puzzlement or wrong answers. Teachers do redesign, but more rarely. They may complain that they lack the time, and thus leave it to researchers who too often fall into the traps mentioned at the start of this chapter. Researchers, like student teachers, design on paper and then have to carry out the plan exactly, observe its consequences, and then go back to the

Getting to Know about Energy in School and Society

drawing-board to produce changes. Indeed this laborious process has achieved the status of a 'model of curriculum development' in educational jargon. Action research has been defined in many different ways (for a thoughtful review see Nixon, 1987), in most cases the objective is to combine feed-back with planning, planning with action, and action with change.

What's Wrong with the Conservation Principle?

In the case of the energy teaching study, there was clearly enough feed-back to stimulate the beginning of planned action. The first simple step was to ask some of the pupils what they thought 'conservation' meant and to compare this with the answer they had given when asked about the Conservation Principle in the written test.

The majority of those who answered that conservation meant 'storing it', 'using it in small amounts', or 'keeping animals in Safari parks', had also got the answer wrong in the physics test. In a way similar to the 'Two Domain' effect discussed in the last chapter, it seems that they had been unable to access the physics meaning of 'conservation' in the face of its more widespread life-world meanings. An interesting sub-group of pupils who came from Greek-speaking homes and who maintained that they had known no other meaning for the word, all got its physics meaning *correct* both while speaking about it and in the test. Without conflicting life-world meanings the physics definition of conservation had been easier to learn and access.

But there may be more problems connected with the form of words than just the alternative meanings. The negative construction of the sentence is itself unhelpful, although it is certainly emphatic. The cognitive psychologist Peter Wason has maintained that negative statements are always more taxing than positive ones and gives the following example to prove it: 'No head injury is too trivial to be ignored'.

My experience of trying out this teaser on long-suffering friends supports Wason's argument. It is not so much that the sentence seems uncertain, but that the very emphasis of the denial — '*no* head injury' (in common, perhaps, with '...*never* created or destroyed') hides the main contention of the phrase. In Wason's example the hidden message is nonsense, so the mind may construct its own sense. In the case of the Conservation principle the positive message should have been quite simple — the total amount of energy does not change — but it is not stated in that way.

Negative statements leave a confusion of possible answers. Stating that one or more possibilities cannot occur, is not equivalent to saying what can. If the energy is neither created nor destroyed, where is it? — in the object? — spread out around it? — in a new invisible form? Theoretically the remaining possibilities are infinite since only two solutions have been eliminated. The mature physicist understands that the Principle is no more than a balance sheet statement about the quantities of energy to be accounted for; but the average 14 or 15 year old cannot easily fathom this out.

The first action in the new teaching programme was to reword the Principle in a more positive and helpful way:

The total number of joules of energy is the same at the end of any process as it was at the beginning.

That was only half the job. Making it clear just what the Principle did mean, opened it up to challenge from the children's own perceptions and understanding of the world. As one thoughtful boy remarked after grabbing me in the school corridor the day after being introduced to the Principle:

'Miss, I have been thinking about what you said and I don't agree. I mean, think of a torch lighting a bulb. Chemical energy turns into electrical energy, right? and then to heat energy in the light bulb, right? Then it all evaporates. So where's the energy then?'

To have a pupil of less than average school achievement reflecting out of school on new ideas derived from the physics lessons, is itself a tribute to the clarity of the new formulation of the Principle. But it also shows that it is ill at ease with common intuitive notions. It did not ring true to this boy, or to many of his friends, and could not by itself form the platform for a new way of teaching.

At a superficial level it was easy enough to see what common sense predicts about energy changes. Such ideas were to be found abundantly in the 'life-world' category answers to the motor car and golfer problems. 'It would', said the pupils about the car or the ball, 'run down', 'come to a halt', 'just stop', or 'run out of energy'. Energy changes do not keep on going, and when they stop, or why they stop, is a result of energy 'disappearing'. Far from being conserved there is a clear pattern in what we see happening around us which suggests that energy is expended — 'used up'. Even the language we use to speak of national energy problems, or of household 'energy consumption', shows that we do not intuitively believe that it is conserved.

The two questions about energy changes — whether it changes in quantity, and what its direction of change might be — are now seen to be inseparable in the learning process. Experience tells us that energy changes take place in a 'running down direction'. This looks like a contradiction of the Conservation Principle, which it is not, and it certainly makes the pupils' application of the Principle much more difficult.

Moving Pupils' Ideas towards a 'Running Down Principle'

The researcher's action at this point would be to interview each of the pupils in depth. That method, made respectable by repeated usage, is based on the assumption that the pupils have clearly defined ideas, that they will be able to articulate them, and that neutral questions can and should be asked. It could just as easily be the case that the pupils have only half-formed views, which they cannot articulate without getting verbal confirmation and encouragement from friends with similarly incomplete views, in the manner of opaque common-sense knowing. Alternatively they might have no ideas at all.

The law concerning what energy changes take place most readily and most efficiently, was a long time arriving in the history of scientific thought. It is still considered to be one of the most difficult principles in physics, and is phrased in a number of different ways which can be shown, with some difficulty, to be mathematically and logically equivalent. Worse still, these formulations are all couched in negative terms. This means that school children are unlikely to be able to articulate any statement of it on their own, and that the existing negative statements may well be confusing, as they were in the case of the Conservation Principle.

In any event the practising teacher's approach is likely to be quite different. She cannot easily interview children, but can ask questions of a class during lessons, and even eavesdrop on talk during informal work. Nor will she labour to ensure that questions are neutral and do not 'lead' the pupil. Teachers' questions are designed to elicit the right answer, if at all possible, because they teach as well as inquire. The teacher is also conditioned from daily contact with pupils to respond to all kinds of feed-back from the class. Put together, this entirely natural approach to teaching and learning problems attempts to understand not only what the pupils already know, but also what paths to new ideas they can follow. It is an active exploration of the pupils' 'proximal learning' positions, as Vygotsky (1979) called them, through teaching, listening and teaching again. The teacher is finding out what the pupils are near to accepting and understanding.

What follows are conclusions based on recorded extracts of pupils' comments to each other and to the teacher as they clustered around the front bench in their school laboratory watching the practical demonstrations that the teacher was performing. These demonstrations were chosen to develop half-formed ideas piece by piece, and so also were the teacher's leading questions. In some cases questions were not needed; in others they were. In still further cases the interrogatory pressure of the teacher produced signs of unease and fatigue in pupils who were clearly *not* ready to understand what was happening. No extra teaching time was available for this work so it had to serve for both research and teaching. The questions probed, but they also stimulated. The succeeding question followed on the answers given in the manner of good teaching and 'reflection-in-action'.

The first demonstration was of a model steam engine heated by methylated spirits which chugged so noisily on the bench that it was often difficult to hear any comments at all! After it was over pupils answered that:

> heat was wasted from the boiler which should have been lagged, and that

heat was lost from the exhaust, and that

this heat warmed the room up a little and even escaped throught the windows to warm the atmosphere or the universe!

and that you could *not* block up the exhaust to prevent the heat from escaping without stopping the engine.

Questioning about the steam pressure and its relation to the working of the engine led, in five out of six classes, to ideas of 'inequality' of pressure being important for the engine to work. In two of the classes there were individuals who jumped from this demonstration to a suggestion that differences, either of pressure or of temperature, were essential and that the engine would run down and stop if they ever came to be the same.

The second demonstration was of a simple siphon formed from a plastic tube filled with water, and two beakers. In each of the three classes that watched this there were several pupils who could argue that it was the difference of level which caused the movement and that this would stop when the levels were the same.

Then came a large and heavy swinging pendulum. In spite of the fact that Galileo had used a form of this to demonstrate the conservation of 'motion', any real pendulum clearly runs down. By now there was consensus in all six classes that the process was 'dying down' and would eventually stop. All classes also identified air resistance as the agent which liberated energy 'uselessly' as something like heat. Three of the classes which had already identified difference and sameness as important factors in the siphon and the steam engine could now answer questions about running down in terms of differences in height which 'made it go' and would 'even out' as it stopped.

The final demonstration was releasing compressed carbon dioxide from a cylinder through a piece of cloth. This was introduced by the teacher speaking about the use of compressed gas to work an air rifle and a pneumatic drill, to connect the demonstration with working machinery. Three of the six classes immediately predicted that some energy would be lost uselessly as heat.

But that was not what happened! Instead, a shower of powdery 'dry ice', was formed — and the pupils rushed closer to touch it. There had been running down, they said, differences in pressure had 'evened out' as the gas escaped, but they all needed help in identifying the kinetic energy of the streaming gas as the 'wasted energy' which they were now getting expert at expecting.

Reformulating the Second Law

From these discussions, using the pupils' own vocabulary, an elementary statement of the Second Law of Thermodynamics, about the direction of energy changes, was formulated.

THE RUNNING DOWN PRINCIPLE
'In all energy changes there is a running down towards sameness in which some of the energy becomes useless.

This is poles apart from the formal statements of this law:

In the neighbourhood of any state of an adiabatically isolated system there are some inaccessible states.
(Not all changes are possible.)
It is impossible to construct an engine which, working in a cycle, will produce no effect other than the extraction of heat and the performance of an equivalent amount of work.
(Engines need a difference of temperature.)
It is impossible to construct any device which, operating in a cycle, will produce no effect other than the transfer of heat from a cooler to a hotter body.
(Refrigeration needs power. Cooling to room temperature does not.)

It will come as no surprise to learn that theoretical physicists tend to prefer the first of these statements (Caratheodory's). It is, after all, more perfectly abstracted containing no references to engines or devices which might be mistaken for real objects, which they are not. In practice all three are negative (*inaccessible, impossible*) and they are also mathematically equivalent.

Practical engineers, of course, cannot be doing with such abstractions. They invent for themselves formulations about the flow of energy through real machinery. They analyze efficiency in terms of 'exergy', 'anergy', or "lost work", knowing full well that only ideal smooth reversible changes take place without energy loss. Some useful work is produced by differences of intensive characteristics such as temperature and pressure, and some useless lost work which it is their task, as engineers, to reduce to the very minimum. At least one of these practical analyses can be shown (Solomon, 1982) to be formally equivalent to the pupils' own statement of the Law.

The Second Law does not figure at all in most sub-16 courses. The standard school textbooks may mention wasted energy but they usually miss out the movement from difference to sameness. All too often they also state that the wasted energy is *always* heat. This is not correct, and a recorded discussion between physics pupils in the Sixth Form both illustrated this common error and then put it right.

> P4. When you change another type of energy into heat it might not all go back into heat, but it's going back into some form of energy.
> P1. The most efficient change is into heat.

P4. Energy into heat energy, but you are not going to lose energy doing it.
P2. Energy cannot be destroyed, lost....
P3. You cannot run out of energy; you can change it all into a less efficient kind of energy.
T. Such as?
P3. You could turn it all into heat....
P4. Where does it get energy in the first place?
P3. It is created with energy. 'Let there be energy!'
P2. But burning is easy.
T. *Is it true that whenever a chemical reaction takes place some of this energy from the bonds is always liberated as heat?*
P1. No
P3. Are you talking about the substance or the net energy?
P4. No, exchanging energy, storing energy.
P1. I was thinking about endothermic reactions.

That answer was a stopper! The misconception, deriving from the familiar operation of friction, had been nailed.

Physicists may find the word 'sameness' unfamiliar in several ways. It was originally generated by a gifted pupil who argued that it described the trend of energy processes better than 'disorder'. A statistical microscopic approach based on the kinetic theory of gases, speaks of order where there is difference, and disorder where there is sameness. There is much virtue in this approach when probability theory can be used to yield mathematical equations. However, when a macroscopic treatment is not possible because the mathematical ways of thinking are not yet developed this 'difference to sameness' formulation proves useful.

Even the absorption of heat in such familiar endothermic processes as the melting of ice and the solution of ammonium salts in water, can be described both macroscopically and microscopically. The processes show 'difference' at first — solid upstanding ice in a flat dish, or crystals separate from water — and then a movement towards 'sameness' sets in. The melting spreads out the substance of water, and the dissolving of the crystal produces a uniformly concentrated solution.

(A session with biology teachers preparing to use the Revised Nuffield A level text showed another use for this formula. Cellular respiration was being discussed and it turned out that some teachers wondered if this flouted the Second Law because more complex biochemical substances had been built up out of simpler ones. With the new formulation, they were all able to identify a temperature difference between sun and plant driving a reaction where anabolic processes took place. They then went on to identify the catabolic processes taking place during respiration as ones in which the formation of carbon dioxide out of sugar and oxygen was a move towards

Getting to Know about Energy in School and Society

Table 6 Golfer's bouncing ball problem (after change in teaching)

Year	Class	Omitted	Life-world††	Storage or reversible	Transformation	Transformation + dissipation
1981	Top ability (25)	9	4	1	2	9
	Mixed ability (27)	3	2	4	5	13
	Middle ability (19)	7	4	4	0	6
1982	Top ability (14)	1	1	0	3	9
	mixed ability (24)	3	2	1	6	12
	Middle ability (18)	4	4	2	7	1
	Total	27	17	11	23	49
	Conservation Principle correct	16	9	9	15	46

(cf. Table 5. p. 128)

sameness from an earlier one of distinction and separation of the substances. They used the same reasoning to explain why a refrigerator working on power from a coal-fired power station, could become cooler inside, and hotter outside.)

Does the Running-down Principle help Pupils?

The final phase in this action research was to encourage the pupils to use the new formulation for solving problems. The best test of how well they could do this seemed to be the same question about the bouncing ball which gave such poor success rates when linked to recall of the Conservation Principle alone.

In the third and final year of research the Running-down Principle was taught *before* the Conservation Principle. Indeed, the latter was simply added on to the notion of running down as a simple balance sheet adjustment — 'but the total number of joules of energy remains the same.' The questions were administered in the same termly test as in previous years. For the problem with the bouncing ball the pupils were asked to state the Running-down Principle, as well as the Principle of Conservation, before answering. The full table of results is to be found in Table 6 above. Its most significant features appear when the results were compared with answers given in the previous years.

- a) 15 per cent, as compared with 11 per cent of the pupils still gave 'life-world' answers.
- b) Storage or reversibility answers were reduced from 37 per cent to 6 per cent.
- c) Simple transformation answers changed little from 37 per cent to 33 per cent.

d) Complete answers including transformations into degraded or wasted forms of energy (such as heating the ground on bouncing) rose from 18 per cent to 46 per cent.

The finding a) suggests that there were plenty of pupils in the final year of research who were no more able than those in the first year with respect to crossing over from the domain of the life-world to that of the domain of scientific knowledge. The other findings demonstrate significant changes in outcome after adoption of the new teaching programme. The very answers that were the focus of dissatisfaction with the previous teaching programme — reversibility and storage — were those most improved by the new one.

Teacher's Action Research

Any classroom teacher claiming that the action she has taken and evaluated in her own lessons proves some substantial pedagogic point will be likely to be met with academic silence. In Britain there is little schism between educational and teaching circles, so that an establishment conspiracy theory is a good deal less tenable here than it would be in some other countries. Nevertheless, it is not hard to show that even the friends of teacher action research from higher education, prize it not so much for the knowledge it produces as for the *virtue* of its practice.

Some three or four years ago, a group of enthusiastic teachers were speaking together about their action research with a lecturer in education — 'What shall we do next?' they said. 'How shall we get recognition for our work?' My advice, 'Publish it in the educational journals', was a very unpopular reply with the lecturer. 'That', she countered angrily, 'would put the cause of action research back ten years.'

Teacher research was seen by this group to be an exercise in self-improvement rather than research. The purpose of classroom exploration was to uncover a contradiction between what the teachers thought they were doing in their classrooms and what they were achieving (Elliott, 1978; Whitehouse, 1989). This was then supposed to lead, through personal trauma, to revised action. A splendid goal, perhaps, but the route to it seemed to be both painful and not a little patronizing.

Teacher action research, as described here, seeks for valid educational knowledge, on a par with other educational research. The serious critics of action research need to question it on this ground and it must be defended in the same spirit.

Can insiders, totally committed to the teaching process, ever have enough distance from the task to validate their explanation for some action research result?

What confidence can there be that other teachers, following the same plan, would achieve the same results? Can a new practice of this kind ever be completely described and replicated?

Validity Analysis

The questions address problems of validity, both internally with regard to the methodology, and externally about replicable findings. They use the language of the researcher rather than that of the teacher, and challenge the findings of the teacher from the point of view of the researcher. Such questioning, based on the thinking of one community but used to examine the claims of another, has rarely managed to produce an agreed verdict in any human controversy, because the knowledge of each community is of a different kind, and based upon different criteria. This brings in the fourth of the factors used to describe the common impasse between teachers and researchers, and stated at the start of this chapter. How different is the knowledge about teaching and learning which is held by teachers and by researchers?

The argument between 'insiders and outsiders' in research is well described in Robert Merton's account (1973) of the immediate post-war inquiry into the position of the Negro in American society. Then the question was whether to appoint a Scandinavian researcher who, as an outsider, might be an objective observer free from the prejudices endemic in a race-torn society; or a Negro scholar who would have personal understanding of all the concomitants of racism. In the event, the first option was chosen and the report was received with much acclaim. It could be argued, however, that it did more to describe than to solve the problems of the American blacks.

In educational research the choice is similar. It is between the objectivity of the outside researcher who might describe classroom interactions with great clarity, and yet not be able to visualize what actions might produce practical solutions; and the teacher whose long-standing familiarity with the classroom and commitment to teaching, might produce tunnel vision and yet relevant suggestions for solutions. In action research it is *the potentiality of action* which needs to be conceived and understood. Conceiving of potentiality always involves imaginative modelling. For solving classroom problems there must be manipulation of a mental model of teaching and learning constructed from the experiences of many incidents. The personal model of the teacher has been built up, almost unconsciously, from myriads of effects in a familiar and relevant context, while the researcher's model is built much more deliberately out of altogether thinner, but more rigorous, research findings in a field of academic study.

Donald Schon puts it in these terms,

The dilemma of rigour or relevance may be dissolved if we can develop an epistemology of practice which places technical problem solving *(such as how to teach about energy)* within a broader context of reflective inquiry, shows how reflection-in-action may be rigorous in its own right, and links the art of practice in uncertainty and uniqueness to the scientist's art of research. (Schon, 1983:69)

This careful phrasing of the problem in terms of *'practice in uncertainty and uniqueness'* picks out the very qualities which are stumbling blocks to any normal claim to validity. Making teacher action research *'rigorous in its own right'* has to accept these concomitents, but transcend them. The action research will have been recorded by one teacher, by methods which lie far away from the usual posture of research neutrality. Could such classroom action by a single practitioner ever have valid generalizable findings for teaching?

The first step is to use the teacher's mental model to create a theoretical underpinning of possible action, using an approach we shall call *validity analysis*. In the present case classroom incidents have begun to model how the children react to the physics concept of energy conservation. Far from action research being the anarchic and pragmatic process that some have assumed, it needs to be continuously and self-consciously critical to a very high degree to achieve analysis validity. Four inter-related questions need to be answered:

- Was there analysis of, and explanation for, what was wrong with the old programme?
- Was the new programme constructed with this analysis and diagnosis in mind?
- During implementation of the action was there continuous reflection on progress with regard to the explanatory analysis?
- Has the new programme been evaluated by comparison with the old programme?

In every case it is only the explicit mental model of teaching and learning, in the special context being considered (e.g. energy conservation), that can answer the questions. The model explains, predicts, and monitors practice against its own interpretation of events.

Anecdotal teacher research in the classroom, substantiated by quotations from the children and good test scores, always has a strong rhetoric of persuasion (e.g. Hustler, *et al.*, 1986; Armstrong, 1983). Too often, however, it fails to show any *'epistemology of practice'* (Schon) through an analysis of what precise problem it was attempting to solve, or on what grounds progress is being evaluated. If all four conditions listed above can be met through a substantiated teaching model, the action of new teaching will be able to claim a logic and rigour of its own. The question of validity is no longer being dodged by claiming that either teaching or reflection is some special art form.

Almost every new teaching programme in science education manages to produce successful outcomes in the hands of enthusiasts (the Hawthorne Effect): this is the reverse of the coin which states that no imposed innovation can be quite teacher-proof. It is also related to the characteristic of uniqueness in reflection-in-action described by Schon. The *validity analysis* approach outlined here, and exemplified in the new teaching programme about conservation and degradation described in this chapter, attempts to meet objections about rigour by means of serious theoretical reflection, rather than by an appeal to skilled know-how. Only if practice is grounded in an explanatory model furnished out of experience and made explicit, through its design, its classroom action, and its continuous evaluation, can the outcome of one teacher's research be generalized to the practice of others.

Chapter 8

The Social Uses of Energy

Education for Society's Citizens

Knowledge for citizenship has been mentioned as one of the three main reasons for teaching about energy. This theme will now be pursued to the point where it involves not only school learning applied to current issues and the students' perceptions of risk, but also personal values, notions of social justice, and political controversy.

The bland phrase 'energy education for the citizen' does less than justice to the radical intentions that often lie behind such words. Malcolm Skilbeck wrote of a prime purpose of education, quite generally, as enabling 'social reconstruction' (1984). Each generation needs to be empowered to refashion the society in which they live so as to confront new moral and physical challenges. Almost always some specialist knowledge will be essential for this process and education will need to deliver this, together with enough civic understanding to effect changes in society, where they are thought to be necessary.

The new philosophers and sociologists of science also want to include social and political aspects of science education because they are anxious that its internal sociology may be taught. David Edge wrote that there are three models of teaching science of which one is 'science as social activity', where 'socio-ethical goals are explicitly incorporated in the conception of science itself' (Edge, 1985:98). Such goals imply that we teach about energy so that the social, moral, and economic aspects of national or international power generation become part and parcel of the physics lesson. It may also imply that the pupils should come to see the making of scientific knowledge itself as a social and fallible activity.

These lines of argument have the widest possible implications. They will affect methods of teaching, the status of the knowledge used, and the sources from which knowledge and information are to be collected. The report of the Royal Society on *The Public Understanding of Science* stated unequivocally that:

Getting to Know about Energy in School and Society

> A proper science education for all must be the starting point for any attempt to achieve a level of public understanding adequate to meet the requirements. (Bodmer, 1985)

It is not altogether clear what should comprise this 'proper' education if the ordinary citizen is to understand the social uses of energy. That will be a major theme of this chapter, but it can be tackled on several levels. The first level of exploration stays close to the normal school curriculum and investigates how students use their school knowledge when thinking about public energy questions. Is the knowledge adequate and relevant in itself? Do students use it? If the knowledge is found to be inadequate for the task, then the curriculum can be altered, in much the same sense as it was in the last chapter when the teaching of conservation was found to be ineffective.

On another level it might be not so much the school learning, but the perceptions of school science and the manner of teaching it which seems to be at fault. The later sections of this chapter will look beyond the confines of formal and informal knowledge, towards affective reactions in circumstances which challenge or even threaten the pupils. Here the results of Skilbeck's programme for education as social reconstruction will begin to take shape inside and outside the science classroom.

'The Energy Crisis' Impact

In 1980–3, when the study of teaching and learning about energy reported so far was in progress, energy had made a dramatic debut as a media topic. The crisis of 1974 which quadrupled the price of petrol in a single year, had filled the nation's screens with 'Domsday syndrome' programmes which introduced the public to the finite nature of our fossil fuel resources, the exponential growth in their use, and the notion of alternative renewable resources. The pre-eminence of the Arab states in the supply of crude oil was suddenly only too evident to the ordinary British citizen who suffered runaway inflation in petrol prices, a sudden lack of plastic carrier-bags, and an increase in the prices of most goods.

Reactions to this crisis would make an interesting study in its own right. Many of the adults interviewed in the first chapter spoke spontaneously about it, and one even reacted to the very first use of the word 'energy' with 'I assume you mean that in a political sense'. In the four discussions on this theme with adults the tentative notes in Table 7 were made:

The Social Uses of Energy

Table 7

Topic	Incomplete understanding	Affective evaluation
Power generation	They use up fuel 'to work their machinery'. We could use electricity instead of fuels. The power from nuclear stations is dangerous.	Fear of pollution and radiation.
Oil (or fuel) shortage	Created by price-fixing.	Xenophobia.
Alternative energy resources	The technology for using these exists already. No technology is needed for 'natural resources'. Sun is the source of life. The sea is an energy source even when calm.	Suppressed by commercial interests 'Natural' is good. Solar power is 'life-giving'. 'But not for good.'
Global perspectives	Conflation with other Third World problems. Can't run out, scientists will invent new resources.	Compassion for the starving, or 'can't think about it'. Technological trust. Gloom 'new problems'.

Time passes and new energy-related issues, new items in the general stock of informal knowledge, and new fears and trusts, take the place of older ones. However, this early small-scale study provided invaluable training for an observer new to a field where the affective components could be so intimately connected with the personal construction of received knowledge. (Some of the items, such as the blind but unhappy trust in scientists' power of invention had also appeared in John Head's work on Sixth Form Arts students' attitudes towards science in 1981.)

The study of adults' views were also used to describe, in broad brush strokes, the life-world knowledge in the community of which the Fourth Year pupils in the main study, were a part. Since none of the adults had studied physics at school, formal knowledge was almost completely missing. However, they displayed plenty of informal knowledge about energy most of which, they claimed, came from watching television programmes and so would be equally accessible to pupils. The gaps and misunderstandings in this knowledge were either responsible for, or a product of, their strong reactions of anger, cynicism, or resignation. If, for example, they believed that alternative energy technologies were already developed, this might well support a conspiracy theory about its suppression by commercial interests. On the other hand, it could be that their existing ideological attitudes about the evil intentions of industry had itself guided a corresponding selection and understanding of information. Attitude and knowledge-reception are often interactive in television viewing (*see* p. 59).

Getting to Know about Energy in School and Society

This mixture of the affective and the informative is a feature of the social stock of knowledge. Nevertheless some organization of informal knowledge was needed for the purposes of research, and this was obtained by examining the frequency with which certain items of informal knowledge turned up in the thinking of classes of pupils not included in the main study. The items were chosen by their intrinsic value for forming opinions about the generation of power. This yielded the following tentative scheme:

Informal knowledge — What needs to be understood?
I   Fossil fuel reserves are running out.
II There are alternative resources.
III Electricity must be generated from an energy resource.
IV Alternative energy technologies are still undeveloped or inefficient.

It is not difficult to see how these knowledge items might link with the Fourth Year formal physics knowledge — storage, transformation, conservation and dissipation of energy. Whether or not the pupils' formal knowledge about energy was used in this supportive way would be a central purpose of this study.

Conservation and Crisis — Talking and Writing

After the pupils had learnt about the transformation, conservation, and (in the case of the intervention years) the dissipation of energy, some teacher-led discussions of the energy crisis were held with each of the three Fourth Year classes, in the three successive years. These were all started with the question 'What is the energy crisis?'.

The pupils' comments were coded for 'Formal (school) knowledge', 'Informal knowledge' (in the above scheme), and 'Affective involvement'. Since relevant informal knowledge had now been defined in just those four items listed above, all other oddments of information (about hydrogen power, how a windmill works, etc.) were coded as evidence of interest, and hence affective. This arbitrary semantic point made the third category something of a catch-all. Contributions of this kind fulfilled a conversational purpose, intellectual or not, outside the main coded categories of formal and informal knowledge about energy generation.

The following example (Table 8) shows how contributions from pupils in three classes in 1981 during a ten-minute class discussion were categorized. These classroom discussions were not reined back, as those reported before had been, by continually repeating the opening question in order to bring the pupils back to the point of discussion. Indeed, it would be hard to know what might constitute 'going off the point' in this case. The discussions lasted for different lengths of time and tended to favour comments made by the more vocal pupils, so not too much weight can be placed on the

The Social Uses of Energy

Table 8

Mixed ability Informal knowledge	Formal knowledge	Affective involvement
Running out of fossil fuel (7)	Used to make electrical energy (2)	We use it up too quickly (1)
Don't know how to use alternatives (2)	There is energy in the oil (1)	We don't think about it enough (1)
We could use wind or solar (2)		We could get energy from splitting water (1)
TOTAL 11	3	3
Middle ability Informal knowledge	Formal knowledge	Affective involvement
Running out of fossil fuels (8)	You can't convert chemical energy back again (1)	Got to save it (1)
		Arabs want more money (3)
		Arabs have plenty of oil (2)
		There's plenty of coal (1)
		Takes time to make coal (1)
		We use it too quickly (1)
		Need a catalyst to make coal (1)
TOTAL 8	1	10
Top ability Informal knowledge	Formal knowledge	Affective involvement
Running out of fossil fuels (7)	We use them for heat energy (2)	So many things we can do with it (1)
Use them to get electricity (2)	Should be no energy crisis if its never lost (1)	We use it too quickly (1)
We could use solar energy (1)	It can't be changed back quickly (1)	Costs too much (1)
	No shortage, energy in another form (1)	Effect on business (1)
TOTAL 10	5	4

numerical count. However, the proportions of different types of comment were of more interest. It was not unexpected to find the number of formal contributions decreasing as the perceived ability of the classes fell. What was more surprising was the dramatic drop in the number of extra items of knowledge, informal and other, given by the more able pupils.

The next sets of data (Table 9) were extracted from paragraphs written about the energy crisis by each of the pupils on their own. These were scored in the same way as above, and then averaged for each class. The same odd feature — less informal knowledge and affective involvement in the top groups was evident once again. In this exercise the pupils recorded their own views without any prompting from others. Such a written test might favour those with greater fluency in writing. It was therefore doubly

145

Getting to Know about Energy in School and Society

Table 9

Year	Group	No. of pupils	Informal average	Formal average	Affective average
1980	Middle	13	1.4	0	1.4
1980	Top	25	1.1	1.3	0.8
1981	Middle	17	1.8	0.3	2.1
1981	Mixed	21	1.7	0.5	1.4
1981	Top	23	1.1	0.9	1.3

odd to find the significant drop in the number of informal and extra contributions from the more able pupils.

Examining the substance of the writing showed that the reason for this failing of the able was connected with the same misunderstanding of the Conservation Principle which was encountered in the last chapter. The top groups spent a large part of their writing struggling to match the Conservation Principle to the crisis. Indeed, one pupil began his writing with the wary comment — 'This is a trick question!' How could there be both conservation and yet a shortage? Some of the 'reversibility' misunderstandings also surfaced here.

> 'The world still has the same amount of energy that it used to, but now it is in different forms to what it used to be. We don't have much chemical energy but more kinetic energy, and we must change it back to the forms we need to overcome the crisis.'

This account from an able and thoughtful child, showed the fruitless effort that was being made to reconcile the Conservation Principle, as he saw it, with the energy crisis. There seemed to be a fundamental inconsistency, and it troubled him. No such qualms, however, worried the less able. The majority of them could also give a correct rote version of the Conservation Principle; but because their new learning had merely been added to the structureless life-world stock knowledge, average ability pupils comfortably ignored all possible inconsistency and gave a sometimes sparkling assortment of informal items of information and affective reaction.

In the third year of the main study when the student version of the Second Law (Running Down) had been taught, and the notion of conservation had been reduced to the addendum 'but the total number of joules remains the same' the paragraphs about the energy crisis were better. As Table 10 shows, the pupils' writing now followed a different, and more expected, pattern. None of the three groups seemed inclined to use much formal school knowledge about energy for describing and giving their own interpretations of the crisis. Indeed, there is little reason why they should have done so. There is no longer a perceived contradiction between conservation and running down. The top group used significantly more of

The Social Uses of Energy

Table 10

Year	Group	No. of pupils	Informal average	Formal average	Affective average
1982	Middle	10	1.4	0	2.5
1982	Top (2)	37	1.9	0.4	1.9

the informal knowledge items which were so valuable for understanding why alternative energy resources could not be just 'switched on'. This was a more satisfying outcome both in terms of the pupils' understanding of the social uses of energy, and for further validation of the new teaching programme described in chapter 7.

Of Feeling and Action

So far no distinction has been made between normative comments about governmental strategy, for example:

'We should do more research into alternative energy', and items of rather esoteric factual interest.
'They could build a hydrogen economy', or expressions of worry and concern.'
'It would be terrible if we ran out.'

Although all three kinds of reaction are needed for citizen participation in decisions about energy generation, they are socially and emotionally quite distinct. The first and third of these, whose main thrust is non-intellectual, are of greater importance for citizens than is the second. Without the spur of feeling, as shown in the third statement, there would be too much apathy for any democratic change. Without suggestions for action, as in the first, such feeling could not be harnessed to doing and would remain as no more than a sense of impending doom.

It is a much harder task to investigate how far any teaching programme facilitates attributes such as *feeling* related to science-based issues, or suggestions for civic *action*.

A comprehensive review of small-scale research in this field in the USA (Iozzi, 1984) has shown almost no effect on the first of these — 'values clarification' — in any of the case studies. The vast majority of all British science programmes studiously ignore feeling and civic action as either outcomes or concomitants of teaching and learning. If the curriculum developers hope by so doing to remove school physics from the arena of politics, they are not only likely to be continually disappointed, but this severe sanitizing of physics could itself prove to be a psychological error of gigantic proportions.

The work of John Head (1981) on personality types and choice of

Getting to Know about Energy in School and Society

physics suggests that adolescent boys who are uneasy with personal decision-making may choose physics because they see it as a subject where authoritarianism rules, and feeling is banished. Physics may serve this clientele well for a time, but it certainly will do little to commend the subject to others including girls (Smithers and Collins, 1984). (Girls who chose to study physics, in the days when it was still an option, did not conform to the 'foreclosure' stereotype which Head identified amongst boy physicists.)

In the research on the energy crisis there were two small pieces of evidence which seemed to attest to this general perception of physics. The first of these was the greater use of evaluative and normative statements by girls than by boys in their writing about the energy crisis. Sentences were first categorized as follows:

Evaluation	*Reportage*
'We must look for other energy sources'	'Alternative kinds of energy are being looked into.'
'They have to be found...'	'Scientists are looking for...'
'I think nuclear power may be dangerous.'	'Methods of disposing of nuclear waste include...'
'People are worried about...'	'They are developing....'

Only later was it shown that those who answered in the first way (evaluation) were mostly female, and those who gave the blander reportage were mostly male. This difference in response does not necessarily impute any lack of feeling to the boys. It is reported as possible evidence of their judgment about *the sort of responses expected in physics*.

The second piece of evidence about the perception of physics as asocial and impersonal was quite fortuitous. At the end of a question which set simple calculations about the heat energy emitted from a power station, pupils were asked: 'How do you think this waste energy might best be used?' The results obtained followed a pattern. In the top ability group 71 per cent succeeded in both the numerical answers, compared with only 38 per cent in the middle ability group. Some were so put out by mathematical question that they omitted them completely (11) and then also omitted the last question about the use of waste energy (9). There was an association between omitting all of the first question and the second of 0.90 (+/− 0.09).

There was a small anomalous group of pupils who managed the numerical parts of the first question and then, curiously, chose to omit the written part on 'wasted energy'. Every one of these pupils was in the top ability group. There is no direct evidence why these more able pupils omitted the easy final rider. They certainly did not lack writing skills, but they could well have had different perceptions of physics. Some at least of the pupils who do well in physics could be those who best absorb the accepted attitudes towards it. All the tests and homeworks set so far in the physics course had called for factual recall, applications of learnt principles, and problem-solving with a uniquely correct answer. Perhaps these suggestible pupils simply discounted a question which began 'How do you think...' as inappropriate for physics, a teacher aberration, and not seriously intended.

Let the Experts Design a Syllabus

Teaching about topics in the public arena has an uncertain history with no consensually accepted syllabus. In a detailed study of radiation and risk in Dutch physics education Eijkelhof (1990) decided to begin by using the Delphi method to establish what physics experts thought that students should learn. No fewer than 63 scientists, with expertise spanning four fields in which radiation is applied — health care, electricity generation, defence, and industry took part in three rounds of consultation. The author used great care to ensure that the expert group should contain 'a diversity of opinion'.

In the earlier discussion of children's ideas about the concept of energy there was no such difference of expert opinion. The scientific community decides upon 'accepted definitions', and the 'consensus view'. Only the educationalists argue together about the best ways of teaching it. It cannot fail to lend a completely different complexion to any study of pupils' opinions if we have to admit from the outset that differences also exist at the most expert levels. Nor is it altogether convincing to calculate a scientific consensus by the method of averages, as is sometimes done in the Delphi method. This prompts the reflection that while voting in elections for regional representatives does take place in this way, the formation of scientific opinion does not. Change in theory has never come about through polling, because it is internal agreement, not the counting of heads, which is the substance of consensus (Ziman, 1968). Calculation of scores and standard deviations based on expert assessment is a world away from the social construction of knowledge.

The celebrated case of leukemia and the Sellafield reprocessing plant provides an instructive example. A large proportion of all 'acknowledged experts' had for years argued that all the evidence pointed against a causative connection between radiation leaks and the cluster of child leukemia cases in the vicinity. The Douglas Black report of 1986 seemed to provide convincing statistical support for this view. The question was certainly not ignored; it was considered, decided, and agreed upon, apart from a very few extreme and often 'less expert' voices. Until 1990 there was even less doubt on this score. But now a fresh report has simply over-turned the earlier views and shown that workers in Sellafield do have a significanty higher than average risk of fathering children with leukemia. No calculus of agreement could possibly have foreseen that switch in consensus, although it is not unique in the history of scientific theory building.

Getting to Know about Energy in School and Society

Delphi studies might also be used to make recommendations about what topics should be included in a physics syllabus, but the question of the order in which they are to be tackled has a logic of its own which is related not to ordering of perceived importance but to the art and psychology of teaching. For many years it has been common to relegate reference to practical application in science to 'the end of the chapter'. The argument was that pupils should first learn theory and concepts in their orthodox decontextualized domain, and then spice this learning with knowledge of related uses in society. The case for placing abstraction after concrete examples has been argued in the last chapter. In the present case there is a second and more pressing reason not to delay teaching about the 'applications' of theory. Placing theoretical abstractions first simply will not do when pupils or public feel at risk, and seek urgently for new understanding.

In Eijkelhof's study the use of the Delphi method generated a list which was headed by learning about 'Background radiation: from the cosmos, rocks, building materials, etc.' (Eijkelhof. op. cit.:37). It is easy to see the rationale for that. Medical applications won the second place, but the emission of radioactive substances after an accident at a power station, and fall-out (but only from the explosion of a weapon!) got third and fifth places respectively. It seems unlikely that, in the immediate aftermath of the Chernobyl incident, the children's interests would have generated a list with the same order of preference. Motivation for study is certainly not enhanced by ignoring the pupils' own interests.

Studies Following the Chernobyl Incident

Eijkelhof and Millar (1988) wrote a report about the non-expert's understanding of radioactivity and ionizing radiation based on a study of the Chernobyl incident as reported in the daily press. They uncovered a common muddle of terms such as radiation and radioactive (substance), and a complete bewilderment over the range of different units employed. In this analysis the emphasis was still on correct knowledge, as it might be in a classroom. Indeed, the authors refer to work on children's alternative notions of concepts in science, for all the world as if there were no difference in kind between understanding life-threatening radiation, and understanding the meaning of 'energy'.

A year after the Chernobyl incident, a British examination board (NEA) set a question in a non-specialist paper (*Science Technology and Society*, NEA) about it for 16+ aged students. The greater part was a comprehension exercise based on an extract from the press about the accident (*Observer*, 25 May 1986). This stated that radioactive materials were pushed 2,000 to 3,000 feet up into the air to form 'a giant radioactive cloud which drifted across Europe'. The examination candidates were asked, 'What effects might this have had?'.

The Social Uses of Energy

Effects on vegetables, on animals, on grass, and on people (cancer) were all mentioned. However, the commonest phrase used was not in the physics vocabulary at all. The students wrote that the cloud would 'poison' the vegetables, animals, grass or people. Was this a gross misunderstanding or just a vivid and appropriate metaphor?

A third study of misunderstandings related to the Chernobyl incident identified a different problem. Robin Millar and Brian Wynne (1988) extracted from newspaper reports an exaggerated expectation of precision in scientific data which led local authorities and schools to use their own Geiger counters in the belief that exact knowledge must always be available. An overly respectful attitude towards scientific knowledge on the part of the public was noted in chapter 5 (p. 85). Excessive trust in scientific measurement is probably another facet of this, which can be particularly unfortunate in times of uncertain risk. The authors claimed it as a serious defect in the public understanding of science along similar lines to David Edge's requirement that aspects of the sociology of knowledge should be taught.

> Understanding the intrinsically limited nature of even the most exquisitely designed scientific experiment or observation has far-reaching implications for a democratic society's integration of science and technology. (Millar and Wynne: 397)

The public's understanding of scientific process was considered in the context of informal knowledge (p. 87) It was shown, using the data collected by Durant and others, that the public's view is context dependent, different for medicine and for metallurgy. Since 43 per cent of the public see science as 'fact gathering' it seems that Millar and Wynne's programme may be very ambitious indeed.

Knowledge and Risk

The preceding sections have shown an uneasy relation between knowledge and affective reaction in reactions to hazardous situations. The school study demonstrated that inappropriate knowledge or perceptions of physics could be a barrier to a sensible deployment of information and social reflection. When students or newspaper reporters face serious potential risk, what knowledge they do have about radiation, probability, or about possible experimental error, often appears to desert them.

Some suggest that an understanding of statistics might mend the situation, but few lay people warm to this. The clichés which compare statistics to 'damned lies' may be a little rough, but there is also very little merit in the dismissive attitudes of the 'hard facts' brigade. In the realms of personal risk 'hardness is' a rare commodity. Calculations need to make assumptions about essentially uncertain conditions, on which there may be rival theore-

151

tical models. Experimental evidence and its 'hard' numerical results are often based on laboratory simulations which cannot match real environmental conditions.

The following passage is by an unrepentant risk statistician, and concerns the dramatic appearance of the hole in the ozone layer over Antarctica, discovered in 1986, which showed that theory and calculation had been sadly deceived. Nothing, however, seems to shake his belief in the comforting hardness of numbers.

i) CFCs are virtually indestructable in the lower atmosphere and thus accumulate.
ii) They are known to be transported eventually to the stratosphere.
iii) Laboratory experiments indicated that chlorine released from the CFCs would reduce ozone.

These were hard facts.... Hard data soon cast doubt upon the theory, or at least suggested counter-balancing changes due to other factors.... (Sprent, 1988)

Those who are concerned about the public's attitude to new technology, like nuclear power, often claim that if only they understood risk statistics and numerical probability better their anxieties would be almost totally allayed. Such views emanate from informed and rational circles where it is assumed that incorrect or distorted statements in the press are the cause, and not the product, of fear or distrust. The Royal Society report on the *Public Understanding of Science* drew special attention to a faulty understanding of probability.

When this point was explored in a public survey by the following low-key question, the public acquitted itself well enough.

Doctors tell a couple that their genetic make-up means that they've got a *one in four* chance of having a child with an inherited illness. Does this mean that:

	Yes	No	Don't know
If they have only 3 children none will have the illness?	4.9	84.2	10.7
If their first child has the illness the next 3 will not?	9.3	80.3	10.3
Each of the couple's children has the same risk	82.1	9.6	8.0
If the first 3 are healthy, the next will have the illness	8.6	80.3	10.9

(Durant *et al.*, 1989)

How people perceive real risk to them is altogether another matter. In the first place the available knowledge has to be communicated, always an uncertain procedure, from source to public. Affective reactions guide the selection and uptake of information, and they also provide an undercurrent of evaluation in terms of the human predicament.

In the aftermath of the Chernobyl incident statistical aspects of risk were a source of confusion to reporters and readers alike. There was a particularly notorious incident when an expert from the National Radiological Protection Board told the press that the very small increase in background radiation 'might add a few tens' to the millions in Britain expected to die of cancer in the next fifty years. A headline in the next morning's Daily Mail read 'A-LEAK MAY KILL SOME BRITONS. DOZENS AT RISK, SAYS SCIENTIST'. Immediately ater this the Department of the Environment effectively muzzled the NRPB and took over the task of dispensing information about radioactivity to the public.

The substantial differences between these two risk situations — one where a hypothetical couple are having genetic counselling, the other where there is a present risk to life — are the immediacy and the personal nature of the threat. These are crucial to the reception and interpretation of information. The question to be asked about perception of risk is not so much what knowledge we have, but what knowledge we pay attention to and trust.

When Eijkelhof summarized the results of studies about pupils' understanding of radiation he commented on the focus of the pupils' knowledge and attention as follows:

> Scientific notions play a small or non-existent part. Reasoning appears to be centred around the perceived risk of radiation for *people*.... Pupils seem to be less concerned with the nature and origin of the radiation. (Eijkelhof op. cit.:87)

Learning through Controversy

In the work which follows this fundamental concern about risk to people, and the controversy it produces, will be explored as a classroom activity. Allowing, or even encouraging controversy about nuclear power within the school classroom will be anathema to some physicists.

> We should also deplore the policy of teaching 'energy studies' as a vehicle for political indoctrination. (Warren, 1986)

It would be possible, although difficult, to teach the items of formal and informal information about energy generation and radiation, to test for and reinforce them, and yet to admit no overt controversy into the classroom.

Getting to Know about Energy in School and Society

Kenneth Baker, when he was Secretary of State for Education, wrote in *Physics Education* that one of the purposes of making science a compulsory core subject was to help young people 'make informed and balanced judgments...about those issues where scientific knowledge is essential' (Baker, 1989). Yet his public utterances showed not a vestige of approval for controversy on public issues within schools.

Arguments for controversy in education fall under three main headings.

1 Civic education.
Our pupils are open to the most sophisticated range of political arguments on the television. They are far from naïve in such matters. Yet civics itself, as a school subject, is more studiously avoided in Britain than in any other European country. Many other educational systems consider such studies not as bringing about indoctrination, but rather as arming students against it.

2 Practice in listening and responding.
With the present emphasis in education on 'communicating skills' it is not surprising to find renewed interest in group discussion. At the very least its advocates mention 'efficient fluency practice, logical skills, and debating skills' (e.g. Ur, 1988). Just taking part in discussion on valuable themes by listening and responding is an active process which opens evaluative social routes to the kind of understanding for which logic, practical work, and mathematics may be considerably less helpful.

3 Teaching in areas with no right answer and hence stimulating personal reflection.
This is the sort of educational objective that inspired the Humanities Curriculum Project (Stenhouse, 1969). It goes beyond the closed agenda of 'taking all the evidence into account' which is still the recipe for less adventurous programmes. It accepts that value judgments will be made which transcend the summation of evidence; hence 'decision-making' — that catch-phrase of the 1980s — cannot but be fallible. As put by Hanley et al., 1970 the aim is:

> ...to legitimize the search; that is to give sanction and support to open-ended discussions where definitive answers to many questions are NOT found (quoted in Stenhouse, 1975)

It is the travelling, rather than the point of arrival, which is of value.

The Humanities Curriculum Project canvassed teacher-directed class discussion, and then wrestled with the ensuing difficulties. Could or should the teacher be neutral in tone? Rudduck (1979) was in no doubt that the strategy of the 'neutral chairman' was correct. In practice many teachers found it difficut or even distasteful. Some moved on to try out the alternative strategy of a 'balanced chairperson' who was prepared to play devil's advocate if only one side of the argument had been presented. Of course, unequal power relations in the classroom mean that no full-blooded disagreement between pupil and teacher can be worked out. What the teacher says does not so much command instant agreement — many 17 year olds seem to disagree almost continually with adult views! — but normal repartee and discussion cannot occur. Social rules preclude the normal exchange and prodding of proferred viewpoints. Only if small groups of pupils could take over the management of their own discussions, at an appropriate point in the lesson, could this problem be diminished.

J.J. Wellington (1986) simplified the general discussion of controversy into two working criteria. A 'worthwhile' controversial issue was, for him, one which involved personal values, so that it could not be settled by evidence alone; it was also one which was considered important by a sizeable number of people. Such apparent objectivity defines the terms but adds little to the arguments about justification. Whether or not objectives such as 'avoiding bias', or 'weighing evidence' (also mentioned by Wellington op. cit.) can be taught in some substantial sense through the consideration of controversies requires proper examination.

The Discussion of Issues in School Science (DISS) Project

The value of small group discussion appeared most strikingly many years ago when my original small Sixth-form group of enthusiasts who were studying *Science Technology and Society* suddenly expanded from 9 to 25. Previously we had spoken together in a group round three joining tables in a small room. They knew each other well and I recall much slouching over the tables, listening casually and intently in turns, making asides to friends which were half-intended to be overheard, and occasional more impassioned passages. The next year I faced 25 fresh students in serried rows across a teaching laboratory from behind the traditionally fixed teacher's desk and podium. The old trap of social space (Harré, 1987; Solomon, 1990) immediately enforced the 'speaking only to neighbours in our own social space' syndrome. I tried a classroom with movable chairs which we arranged in a giant circle. Each student now stared across a diameter of some 3–4 metres at a distant peer in solemn silence. It was too far to talk, they could not shout while I was there, and no one felt like making a speech. Eventually we returned to the first laboratory and simply legitimated small group discussion between neighbours. Somewhat to my surprise it worked extremely well.

These kinds of discussion were put to research use in the DISS (Discussion of Issues in School Science) project. This began in 1988 as a part of the linked programme of projects on the public Understanding of Science. In the Royal Society report on this subject there had been a recommendation to set up 'science indicators', of the kind produced yearly in the USA, to

Getting to Know about Energy in School and Society

ascertain what the public knew, what they were interested in, and how well disposed they felt towards science. Fortunately the programme managed by the Science Policy Support Group with funding from the ESRC had a wider perspective on what might be included. It soon appeared that none of the words 'public', 'understanding', or even 'science', was neatly circumscribed. The research projects had to include groups of the 'public' who had special needs to understand, 'understandings' which were composed as much of affective and social factors as strictly cognitive ones, and 'science' as seen by communicators and those who received and reconstructed the communications.

Discussion usually needs a trigger. There was a considerable body of research from the USA which indicated that most students attributed the greater part of their knowledge of environmental issues (this included the impact of nuclear power) to television (Weiesenmayer *et al.*, 1984). There was no reason to believe that it would be any different for British students. Another interesting point turned up in communications research literature. Those who had watched interesting programmes often felt the need to discuss them afterwards in order to 'know what they thought about it' (e.g. Blumer *et al.*, 1979; Hodge and Tripp, 1986). Showing a short piece of television to begin discussion became part of our research methodology.

Small group discussions, with 3 or 4 students controlling their own tape-recorders, were held in ten different schools, in each of the two school years of the project. Most of the students were of only average ability, many staying on one year in the Sixth form merely to improve their tally of passable examination results. All were in classes studying *Science Technology and Society* (NEA) for the GCSE examination. This is a course which lasts one year and sets out to cover some basic aspects of the subject such as the changing nature of scientific knowledge, and democratic decision-making processes; some terms and concepts such as energy and nuclear power; and some more detailed optional topics of which one was energy. Most important of all was the requirement for small group discussion work.

The students involved in the DISS project held discussions on six different themes during the year. The second of these was about nuclear power. It is from some 35 small group discussions, some lasting no more than three minutes, some going on for 30, that our illustrative material will be drawn.

In all the classes the teacher had first shown their students the same edited video excerpt from a television programme on nuclear power first screened shortly after the Chernobyl incident. It began with some instruction on how a conventional gas-cooled nuclear power station worked. It showed how different nuclear policies operated in Britain, France, Sweden and the USA, and the public reactions to nuclear power in each country. Both the Three Mile Island incident and the Chernobyl disaster were reported and discussed. In the final minutes of this 20-minute video there was a heated confrontation between Lord Marshall (then Chairman of the

CEGB) and Jonathan Porritt (then Director of Friends of the Earth). As the excerpt ended abruptly — almost in mid-speech — there could be no escaping the controversial nature of the issue. The third person in the debate was an American risk-assessor, so that aspect of nuclear power was also highlighted. The teachers set no task beyond the comment that 'there is plenty to discuss in that'; then the students switched on their recorders and began speaking to each other.

Students' Assessment of Risk

There is a considerable literature about risk perception, of which very little is empirical and still less derives from discussion rather than questionnaire or survey. Using data gathered from the DISS project it is possible to do more than simply rehearse the different views. The transcripts supply exemplar material of how students construct their perceptions of risk, and show how much weight is given to different types of risk argument.

One of the most striking features was the way in which the students tried to increase the saliency of the issue by questioning each other about what their reaction would be. Whenever a decision had to be made, the students were quick to personalize it. This brought it a little closer to home, and almost never failed to obtain a strong response.

> 'What would *you* think about having a nuclear power station in *your* town?'
> 'Would *you*, wouldn't *you* mind if there was a power station by *your* home?'
> 'I mean, but what if it was here? You wouldn't be saying that, would you, if it was right in *your* back garden?'

Even when actual numerical data on risk were available the way in which students reacted was little different. The probability was weighed up, as just one more piece of evidence along with others. Far from considering it 'hard' and unimpeachable, these discussions showed that mathematical data had to be personalized in order to be found worthy of trust, like all the other information.

In the video the American risk assessor had given an estimate that there was a 45 per cent chance of a nuclear accident happening once in every 20 years. Almost half the discussions referred to this potentially alarming statistic, but the students coped with it in different ways. Some used the personalizing strategy:

> P1. But don't you think it's a bit staggering that there's a 45 per cent chance of the same thing happening in Three-Mile-Island again in the next 20 years? Don't you think that's awful?

> P2. And just think, it's gonna happen to one nuclear power station isn't it, and we've only got 16, and I wonder how many there are in the world — because then you can work it out, the chances of it happening in Britain, can't you, if there's going to be one of them melt-downs and we've only got 16 in the country?
> P3. Yea, but what would you do if it was in India, you know, all those people.
> P2. Yea. Are there any power stations right next to us?

In other cases it was clear that different students paid attention to, and put their trust in, different facets of the problem, and different types of reassurance.

> P1. A 45 per cent chance that there's an accident...
> P2. ...in the next 20 years.
> P3. In the next 20 years, so that's pretty good isn't it?
> P2. Pretty good.
> P1. I think it is pretty frightening!
> P3. Or pretty frightening, yea, but they're taking procedures aren't they, and they've got to spend more money on that then.
> P1. But there is still a 45 per cent chance of there being an accident.
> P3. ...but then they are doing their best aren't they? They put a lot of money towards it.

Another couple of numerical estimates were given by Lord Marshall. One of these was that 'nothing in the whole world' can be 100 per cent or completely safe. This was quoted by many of the students with great approval. It seemed to have a ring of proverbial wisdom about it, by contrast with what scientists were thought to say usually — 'They blind you with figures!' or 'They should learn to speak English!'

Lord Marshall's second statistic was treated more cavalierly. He had claimed that there was a chance of no more than one in ten million of the new Sizewell B reactor blowing up. This almost unimaginably high figure was reduced to comfortable proportions in an effort to relate it to a personalized context — 'living next door'.

> 'Yea, the biggest problem with nuclear power is just the accidents. It's a million, I mean if you live next door, it's a million to one chance there's going to be a major accident. A million to one chance — that means nine times out of ten you're safe.'

These excerpts highlight not so much an inability to appreciate probabilities, as the twin processes of *constructing* an understanding of risk by:

a) paying attention to those aspects of probabilities which they found important, and
b) reformulating them so that they felt comfortable and familiar.

Curiously enough the phrase 'to feel at home with' science was used both by Collins and Bodmer (1986), and Baker (1988) as objectives for citizen understanding of scientific knowledge. From the main body of their writing, however, it seems that they were using these words without any suggestion that scientific knowledge would itself be transformed in the process of lay understanding. But the reception of knowledge about energy in every chapter throughout this book has been shown to be more or less of a reconstruction. It is hardly surprising therefore that, in the context of risk to the person, this process should turn out to be a yet more radical re-making of the information received, whether couched in words or in numbers.

Weighing Risk and Benefit

It is thought by some that the advantage of quantifying risk is that it provides a basis for risk/benefit or, as it is more often called, cost/benefit analysis (CBA). Those whose task it is to look into the risks involved in locating large new complexes, for example the siting of London's third airport, may be instructed to put a figure on every deleterious effect which might be expected, from the disturbed sleep of local residents to the threat to nesting birds on an estuary. Conclusions would arise, it is supposed, out of the summation of figures. Naturally enough this task is almost insuperably difficult in cases where people or the environment are at risk. In simpler circumstances, such as tooling up for a new industrial process, CBA is widely used.

The problem with quantifying risk is not so much that it cannot be done, although with tolerances which may have to be wide, and with hidden errors that only later research will reveal, but that it will not be universally acceptable. Perceptions of risk are personal, context-dependent, and almost incommensurable. Where hidden risk is inflicted upon us by others we find it intolerable even though it may amount to only a very small chance of harm. When we voluntarily take up a risky sport, such as rock-climbing, we may not only accept but even welcome the risk. If the risky event is under our control, or if the effects of it may be treated and reversed, we both consciously tolerate higher risk and also perceive it as being less hazardous than others do. Some years ago an official of the CEGB tried to allay the public's fear by stating that the risk involved in living near a nuclear power station was no more than in driving an extra three miles on a motorway. If this did not soothe the public, and obviously no one can know whether it did or not, it will not impress those who reckon to avoid road accidents by driving skilfully. Who can 'live skilfully' near a power station?

Getting to Know about Energy in School and Society

Nevertheless, estimating the severity of risk is at the heart of all kinds of evaluation. Indeed, evaluation is essentially 'a weighing up' process in more or less complex situations. In one type of case there are both benefits and risks to a certain policy, for example in extending a nuclear power programme. In other cases people can see risks in either following, or in not following, a certain course of action. The students discussed what would happen in terms of a shortage of electricity and jobs if we did not have nuclear power generation. There were also plenty of discussions where the risks of nuclear and of fossil fuel power stations were weighed up against each other.

Whole pages of discussion transcripts could be presented to illustrate how well these average ability students understood the nature of the balancing and evaluating process. Indeed, in the minority of disussions where there was no exploration of risk, counter risk, and benefit, not only were the discussions short but they also gave the impression of involving no personal reflection on the information given. 'Right, we are all agreed then', and 'There is nothing to talk about, we are all against (or for) nuclear power', prevented all productive discussion. Far more often the students considered one factor after another, saying: 'Yea, but it's cheaper', 'But you cannot rely on (other resources)', 'What about unemployment though?', and more judiciously 'You've got to balance it', 'there are good points and bad'.

Some students, however, did seem to need more time to explore and evaluate than did others.

> P1. And if there is an accident it costs a lot of money to clear up, and the effects are felt years and years after.
> P2. So you're changing your mind again now?
> P1. No, no.
> P2. Are you in favour or not?
> P1. I'm, discussion, I'm discussing...
> P2. All right.
> P1. You've got to look at all the different points and argue it out, haven't you?
> P2. ...before you come to a conclusion.

Scientific and Social Knowledge

As in the earlier study about the energy crisis there was common knowledge about supplies of fossil fuels running out. These older students knew more about alternative energy technologies together with some estimate of their efficiencies. Many had been studying aspects of energy policy in their STS course and some knew about pumped storage hydroelectric plant, tidal barrages, or fast-breeder reactors. However, it was amusing to find, amid all this sophisticated knowledge, occasional remnants of the old trust in energy from the sun which we found in the youngest pupils' reactions to energy, still persisting. *'Yea it's (solar power) safe. No one's gonna be hurt by the sun!'* The hugeness of modern designs of windmills which would deliver only a fraction of the energy from a nuclear power plant, as well as the poor harvest of energy from wood (the video showed both a wind energy farm and willow trees being planted as an energy source) were both the subject of quite frequent critical comment.

For these young people just about to leave school, the risk of unemployment probably ranked a good deal higher than it might have done for other groups of the public. Here again the students often questioned each other, making the points as tough and intimate as they could in order to probe, persuade and decide. It was not surprising to find that boys were more likely to consider working in a power station than were girls. The following extract is taken from the transcript of a discussion between three boys.

P1. I'm not going to work in a nuclear power station.
P2. Right, what happens, right, if you haven't got a job and someone asked you if you would do the job and you was unemployed for about six years?
P3. Of course you would. You would go and work there.
P2. You would, you would work there.
P1. Would you?
P3. I would.
P2. I wouldn't work there. No.
P3. The reason why I would work there is that it's only very rarely that mistakes happen.
P2. Yea, but the pressure there...
P1. But on the programme it says 45 per cent chance of it happening again in the next 20 years.
P2. And look at the pressure you've got. You've got gadgets all over the place. You press one and you miss it and press another — a leak!

That extract, with its striking use of social and psychological insight, exposes an aspect of risk assessment which is often ignored in more traditional analyses. Empathic re-enactment is a valuable tool for refining understanding.

Using Empathy in the Reception of Energy Knowledge

Our 17-year-old students have been learning about people for far longer than they have been learning about energy. Children of no more than 10 or 11 are able to make subtle predictions about the way those they know may react and what they mean when they use certain words. When a child comments on her mother's remarks about information on the television, 'that is just what I would expect you to say!', she is reflecting upon knowledge from a personal perspective which is not her own. This involves a modelling of another's point of view — as complex a task in its own way as any scientific thinking might demand in its domain of abstract thinking. In the social world such 'exchange of perspectives' (Mead, 1934) is central to all meaningful interactions. No doubt our students in the DISS project found this kind of movement from the symbolic to the everyday domain easier for the construction of people's meaning than for physics, because of their much longer familiarity with it.

This modelling of perspectives on information demands a fairly detailed knowledge of the other person. Without this the listener can do no more than build up a caricature of the person and use that. In the DISS transcripts there were plenty of examples of this (frequently scurrilous) process of interpreting another person's perspective. They said, 'Well he would say that', 'Did you see he went red in the face?, so he must have been lying', 'He's an idiot, he's always against nuclear power', 'The Americans are paranoid'. These comments often seemed shallow, prejudicial, x-enophobic, and calculated to denigrate points of view which had already been rejected by the speaker. In just a few cases, such as the common perception of Sweden as being a very 'environmentally good' country, the knowledge given (that it would be possible to phase out nuclear power and use only renewable resources) acquired a slightly greater measure of reliability, because of the students' construction of the Swedish character.

Imputing bias to figures in the videos was common in all the discussions. Even when two scientists of apparently equal standing and probity disagreed, caricatures from the social world, rather than uncertainties in scientific knowledge, were used to resolve them — a point which is similar to that of Millar and Wynne in their discussion of imputed scientific certainty. Scientists were thought to be biased or 'bought' just as often as were politicians, in all these discussions. This seems at first sight to be at odds with results calculated from data deriving from a large questionnaire study (Breakwell, 1990). However, active construction of received knowledge may be a surer, or at least a different, way to probe for 'scientific cynicism' than is the ticking of boxes of the 'Who would you trust most...?' type.

This personality factor in the reception of knowledge also came to light in a study of the exchange of information between Cumbrian hill farmers and MAFF radiation scientists in the aftermath of the Chernobyl incident. Wynne (1988) reported that after the 'Whitehall' scientists had stayed overnight on the farms for the purpose of monitoring sheep, they and the farmers began to understand each other better. It seemed as if having breakfast with a knowledge-giver enabled the other to begin to construct, in some small way, both his character and perspective. Thereafter messages from London about necessary radiation precautions were more accessible to the farmers.

Empathy and Social Justice

Empathy with others played two other important roles in these discussions. In the first place it was put to good effect by the students for understanding each other. Indeed, it became almost commonplace for members of the research team to suspend their own attempts to understand rather inarticulate passages, in favour of reconstructing them from the other students' responses. These groups of friends not only understood each other well — far better than we did — but they also used their powers of empathy to complete each other's sentences.

The following extract came after about 5 minutes discussing risk, including the 45 per cent chance of an explosion in the next 20 years. There was a short pause, then P1 (a boy) began again with an attempt to see what it might be like in Russia and draws the others with him.

P1. I mean I don't want to, I mean, even think about the people who live next to Chernobyl.
P2. (Quietly) Yea.
P1. I mean that is a bit heavy duty, that is.
P3. Yea, it is actually.
P2. I mean I wouldn't feel safe. I wouldn't want my kids to grow up with it.
P1. Yea.

These moves, in which empathy is used to think about those who may be suffering, is only one short step away from identifying groups within society who are particularly at risk. This is where the study of risk from the generation of power ceases to be about the reception and interpretation of knowledge about energy in all its manifestations, and becomes instead a reflection on social justice and possible action.

David Bridges, in his reflections on *Education, Democracy and Discussion* (1979), identified a number of stages and objectives in this sort of sharing of perspectives. Roughly translated these become — clarifying one's own views and values, understanding a variety of responses for which empathy is an invaluable tool, seeing these as a basis for choice between different or complementary value positions, and seeking possible strategies which will resolve conflicts. John Rawls, in his influential *A Theory of Justice*, laid great emphasis on thinking about the rights of others from behind 'a veil of ignorance' about one's own situation. This requires the use of both selflessness and empathy.

In what follows we shall simply be listening to the students build upon their constructions of groups at risk. Even the youngest pupils have a sense of what is 'fair' or 'unfair'. As they grow older some sense of the 'rights' of people should stiffen this intuitive feeling. It should also grow wider to embrace other sections of society, other nations, and even generations yet to

Getting to Know about Energy in School and Society

be born. When our students discuss a topical national issue we would hope that they can not only use and apply these developing values, but also harness them to a sense of civic awareness, knowledge and responsibility. Out of this should grow suggestions for action. Of course, finding new and effective civic strategies is bound to be a tall order for such young citizens in the case of nuclear power whose pros and cons have been debated so widely and so long. But the effort alone will signal some success in citizen education.

> P1. Yea, but coal isn't going to run out straight away is it? It's going to go out gradually so people are still going to be with coal when....
> P2. People will have coal in their houses for quite a while, about 70 years.
> P3. Yea. We will be dead and gone when...oh sorry.
> P2. You shouldn't take that attitude.
> P1. I know I shouldn't.
> P3. Think about our kiddies.
> P1. Yea I know, except I'm not having children....

Sometimes a sober balancing of risks from different strategies was interrupted by an act of empathy for people.

> P1. If they shut down nuclear power stations, everybody who was working there, right, there would be so many people unemployed.
> P2. Yea, but on the other hand if they do it the other way (keeping the power stations but the workers go on strike for greater safety) there's going to be so many people out of their job and everything, you know.
> P3. Is there houses near power stations?
> P1. Yea. There's loads of houses near power stations.
> P3. They shouldn't build them near power stations. They should be moved....
> P2. Buildings, everything.
> P1. They shouldn't build them near houses because...
> P3. ...because of the people...
> P1. ...like even if they say that it won't affect their health, what guarantee have they got that it won't?
> P3. None.
> P1. They've got no guarantee on that.

And, as usual, strategies were worked out by personalizing them and questioning each other.

P1. Electricity would be more expensive.
P2. Would you mind paying more taxes?
P1. No, because it would be safer for people. I wouldn't mind.
P3. I think it's really selfish actually because if anything happened it wouldn't be just this country that would be affected, it's all the countries of the world.

What is Achieved?

This chapter set out to examine how, and indeed if, students could use their knowledge of energy in the discussion of social issues. On one level all the studies show that the students can and do, just so long as there is no clash between a misunderstanding of the energy teaching in school and the public knowledge about issues. The benefits accruing from such activities, however, are mixed and not altogether as some educationalists seem to have expected.

One of the rewards most often promised has been 'debating skills'. Even as this book is being written plans have been announced for a new and expensive city technical college to be built in London's dockland, dedicated to a study of 'citizenship'. Once again speeches promise such educational goods as understanding the rules of society, acquiring a body of knowledge, developing and exercising skills such as debating, taking part in elections, playing as a team member, and learning democratic behaviour. These skills are not new to educational rhetoric. Indeed, they were the substance of Arnold's aims for the public school of the nineteenth century.

What has been heard in the snippets of discussion reported here does not include debating, or playing as a team. That would have implied skilled verbal tactics, and the formation of a collective view to be used in opposition to those of others. These groups of students supported each other, providing help so that a variety of views could be expressed. And when the discussion was already some days, or even a week, old, those students who were asked to do so could usually write quite detailed reports on what they and others had said. Often too, the individual's written report showed that ideas had not been prematurely closed down as simplistic 'decision-making', which mock democratic processes forcing a tally of heads may often bring about. Instead, the students had gone on reflecting about the issue and the personal values involved.

Personal values and concepts of social justice are rare visitors to energy lessons in the school laboratory. Physicists may feel uneasy about teaching anything resembling ethics, and have been heard to suggest that this needs specialist teaching in classes on personal and social education. I doubt very much if this would be a valuable outcome. It could further strengthen a kind of disciplinary apartheid in which physicists teach 'valid science' while others

Getting to Know about Energy in School and Society

are left to cope with the 'soft' side — views and values. Ziman (1980) wrote about this outcome that:

> ...it reinforces, without question or comment, the widespread sentiment that science should be the only authority for belief and the only criterion for action.

What is even worse, it strengthens the belief that scientists, and their school proxies, the science teachers, put their trust in something called 'hard facts', and do not care to reflect on those deep personal values which enrich all human thought.

Chapter 9

The Defining Technologies of Energy

Technology in Education

The last chapter included reference to the technology of nuclear power production for a purpose which was more closely related to ethical education than to engineering. No study of how we get to know about energy can possibly be complete without technology: indeed to have reached the final chapter of this book without a substantial discussion of it, is itself symptomatic of a pathology in education. David Layton (1984) has examined the history of the split between science and technology education only to conclude that it is at least as old as state schooling.

Energy is studied in both 'pure' physics and 'applied' technology or engineering. It is central to both, but is it the same in each setting? This question is intended in a didactic and a psychological sense. To dismiss it with a comment about conceptual or procedural knowledge is to miss a central point in the definition of a school discipline.

Technology arrived at its present position of high profile at precisely that moment when the crafts — metal and woodwork — lost their more modest but previous secure positions. No Minister of Education would have made either of them compulsory. In the early 1980s technology staked its claim to the status of an important school discipline on the grounds of its cultural and theoretical standing. The evidence which won the case was related to current concerns about economic power and monetary ideologies; it was the TVEI initiative from the Department of Trade and Industry not the Department of Education and Science which helped to raise technology to the level of a core subject in the new National Curriculum, above modern languages, religious education or history. In previous ages such an 'intrusion' of a practical subject would not have been tolerated. Layton refers to this as the 'rehabilitation of the practical' but it is debatable whether technology or engineering have ever before achieved so respected a position. (Even when Victorian industry launched the Great Exhibition of 1851 and built the Imperial College of Science and Technology there was strong and successful opposition to technological education in schools.)

Getting to Know about Energy in School and Society

One reason for the astonishingly rapid promotion of school technology may be the peculiar sensitivity of this subject to local culture. In a geographical sense this leads to what has been called technological style (Bijker, Hughs and Pinch, 1989) where power stations and substations curiously reflect how a local community lives, rather than entirely imposing new lifestyles upon the old. It is one of the central tasks of all technology, and especially those of energy, to respond to the local complexion of need. In that sense the school advocates of technology were entirely correct in their support for its cultural status. At a time when every news bulletin on the nightly television contains references to industrial production and indices of inflation, technology, whether in school or college, will include economic awareness to match. That is the measure of its adaptability.

This chameleon quality means that technology may shift to suit a variety of pupil interests — baby-care, structural engineering, or electrical generation. Science is far from unrelated to ways of living, its images and concepts are culturally dependent, but it has nothing like the flexibility of technology. In order to understand the contribution of technology to learning about energy we need to understand this interaction between technology and culture in a series of historical cameos. There have been references to the history of scientific thought throughout this book. The rationale for them was not that children evolve through ancient ideas in some genetic sense, but that the ideas have left imprints in the language which we all use to speak about energy. The same is true of technological images. For lay people the changeover is not only slow because of the ponderous movement of the school curriculum, it often involves two images of power, the older and the newer, existing side by side for many years. In the comfortable attitudes of the life-world, inconsistencies are so easily accommodated that, even after the interval of a century or more, vestiges of the older way of thinking may still be evoked by a motto or figure of speech, almost without a jolt.

In this chapter the topic is technology within commonsense thinking, rather than scientific explanation. The related theme of 'knowledge from doing' was explored in chapter 5 because the knowing associated with personal action makes links with structureless informal knowledge. The special problem mentioned there, the tacitness of everyday doing, is less prominent in technology. Here the artefacts are concrete, often they can be seen working on a daily basis, and their potentiality for forming analogies to other processes are immediately obvious.

A Defining Technology

This term is taken from David Bolter's excellent book, *Turing's Man*. His thesis, if I have understood it correctly, is that the dominant technology of

The Defining Technologies of Energy

every age provides not only the means to improve the material lot of people, it also gives them a creative metaphor. Weaving, for example, was not only a daily occupation in ancient Greece, it also gave the people a powerful way of thinking about the Fates, and about the threads which pulled at 'the fabric of their lives', a phrase we still use today. By Bolter's thesis the Renaissance becomes not just a period in history when new ideas were suddenly and inexplicably rife; it was a time when the mechanical clock gave people both a new way to think about machinery and, at the same time, the vision and ambition to make accurate measurements of time and movement in the universe. So the clock technology fathered all the creativity which went into making the new astronomy of Galileo, Newton and Kepler, as well as an internal vision of how it might be happening.

This is dramatically illustrated in our age by the radically new understandings we have gained through the use of our own innovatory technology. The computer has shown us 'real time' and 'computer time' where memory is no longer a reflective, imaginative reconstruction of how we remember things past, but a kind of accuracy in limbo, stored, filed, and out of contact with the current file. Of course it is true that the older meaning of memory has not been obliterated, just shifted sideways to be used for different occasions. This addition to our ways of thinking has happened for most of us in little more than a decade, a mere twinkling of an eye compared with the changes brought about by earlier defining technologies. Essentially, however, it is as mind-blowing and mind-forming as any which have gone before. From the potter's wheel to the Turing machine, technologies have been for humankind every bit as much tools for understanding and imagining, as for making and doing.

The historian searches for the exact moment and cluster of circumstances which brought some new technology into being, and then explores other details of innovation in its mechanisms. What we need for examining the defining technologies of energy is not so much the dating of their introduction as a feel for their domination in the cultural imagination. It will be the importance and, above all, the familiarity of the device which will lead people to think in a certain way.

At a slight remove from this everyday influence is the conscious use of the workings of technology by innovative thinkers as an available metaphor to grope for a way to describe subtle and half-formed concepts. As Bolter wrote:

> Almost every sort of philosopher, theologian, or poet has needed an analogy on the human scale to clarify his or her ideas. (10)

And since every technology requires a power to drive it, we are likely to find ideas about energy changing with the defining technology of the age.

No Prime Mover

The earliest technology in every culture, as far as we know, was the powerful astronomical/astrological complex. The physical remains of this technology — the stone circles of ancient Britain or the Ziggurats of Babylonia — are little more than relics to their chronological purpose. However, there are also associated pits with cremated human bones at Stonehenge, and a substantial library of clay tablets from the astrological craftsmen of the Middle East, to show that the technology was intended to make a more substantial contribution to human welfare than any modern calendar could do. Telling the time and the seasons was useful enough for agricultural communities, but foretelling disasters or ensuring a protected life in the long hereafter must be considered of far greater value still. Indeed, it is doubtful whether any succeeding technology has ever claimed so much in the sense of long-term utility, or short-term wealth-creation.

The effect on everyday thinking about movement and energy was very powerful, as was its impact upon ancient religion and philosophy. Their first and most pervasive message was the complete effortlessness of circular motion. The stars that revolved nightly round the Pole Star seemed to teach that force and energy were eternal and unchanging. On earth no such unchanging and effortless motion existed. So what the skies presented was an illuminated metaphor for thinking about perfection in movement. Inevitably this was read and used in different ways for different purposes.

The great Chinese 'Book of Changes' begins: 'The movement of heaven is full of power...strong and unresting.' The prayer-wheels of Tibet pick up some of the religious force of this idea. It was movement in a circle, and not rest, which was the perfect state of things unsullied by opposing force. According to Needham (1962, Vol. IV) there is little in ancient Chinese science about movement, but the notion permeated Tao thought as 'ceaseless motion in cycles'.

The Greek philosophers also wrestled with the idea of perfect movement in circles and compared it to another common technology, that of spinning. In their thinking, perfection was linked to logical necessity and also to mathematical and musical harmony. Once again there is no felt need for any energy to keep things revolving, but some inkling of that mathematization of the universe which was to be their legacy to science. Plato wrote of heavens that:

> ...the spindle turns round on the knees of Necessity. On every one of its circles is a siren who travels round with the circle and cries out on one note, and these eight notes are joined into one harmony. (*Republic*)

Nearly two thousand years later Dante and Kepler were direct heirs to this imagery about celestial movement and music. Spheres of perfect and invisible crystal carried the planetary spheres in endless revolutions round the earth, but with just one essential change. Now the motion had to be kept going — by an input of divine power for Dante, and by magnetism for Kepler. By this time new technology had brought about revolving machinery for performing daily tasks so, as we shall see, continuous motion without an external force to keep it going was less easy to accept. All the crystal spheres of the perfect heavenly machinery were geared to the outermost one which was turned by power from the Almighty. A *primum mobile*, a source of energy, was essential even to the cog-wheels of heaven.

As the Scientific Renaissance blossomed, this notion that the heavens were no different from the earth, as far as physical principles were concerned, gained strength. Eventually Isaac Newton wrote it into the second volume of his great *Principia* — the *System of the World* — as one of his basic Rules of Reasoning in Philosophy. Movement could no more be effortless in the heavens than it could on earth.

Powered Wheels

Energy in the natural world was sometimes dangerous, often powerful, and always mysterious. In the earliest days the energy of rivers in spate, of hurricanes and of volcanoes were simply attributed to spirits or gods. Slowly people learnt to use some of this power to turn the wheels of their machinery so that laborious tasks could be performed for them. The first of such tasks were probably irrigation and the grinding of corn. The first wheels turned by natural power were worked by water.

Waterwheels may have come from Egypt, from Jutland, from China or from Rome. Some were on a vertical axis, and some horizontal; in some the water moved the wheel from underneath, in others the path of the water was carefully engineered so that it arrived, by chute or aqueduct so as to deliver power against the blades at the top of the wheel. This had the advantage that by falling down with the turning of the wheel the water could deliver its gravitational energy as well as its kinetic energy. This final stage of technology, the over-shot wheel, was powerful enough to grind the corn for a large village. By the time the Domesday Book was written it seems that almost every village in England which had a suitable river, also had a water-driven mill.

The history of the windmill is more obscure. The ancient world did not use them and yet, when records take up the story of common technologies again in the tenth and eleventh centuries, the windmill is there, in northern Europe, a common sight and a creaking sound on exposed sites in many countries. They were particularly common in the north where winters were severe enough to freeze rivers and so render waterwheels less reliable. This goes some way towards explaining why Don Quixote was amazed enough to charge at the flailing arms of a windmill. In Cervantes' time, the late-sixteenth century, windmills were still something of a rarity in central and southern Spain.

The world of the Renaissance, of Francis Bacon and Rene Descartes, was enormously proud of these achievements, which were seen as a step in the great human project of subduing Nature and using her forces for their own ends. The optimism and even arrogance of their claims may jar upon our more apprehensive approach to nature.

> We have useful knowledge by which, cognizant of the force and actions of fire, water, air, the stars, the heavens, and all other bodies which surround us — knowing them as distinctly as we know the various crafts of artisans — we may be able to apply them in the same fashion to every use to which they are suited, and thus make ourselves masters and possessors of nature. (Descartes, 1637)

For country people in England, waterwheels and windmills were familiar enough by the fourteenth century to enter into the popular culture as images of energy, in two distinct senses. First, they really did deliver a great deal of useful power. Waterwheels were used to drive saws, bellows, and fulling machines, as well as for grinding corn. Indeed it was water-driven machinery, and not steam engines, which launched the Industrial Revolution by working silk and cotton mills by day and night. They were also used to turn cranked rods which worked simple bellows to pump air into the early blast furnaces. These then turned out the pig-iron which would eventually be used to fashion the engines which supplanted waterwheels. Engineers and craftsmen built the gears and cogwheels which delivered the energy, changing a fast and fitful flow of water into the large steady force behind the slowly turning wheels. This almost inexorable power could then be likened to the inexorable 'mills of God' which 'ground so exceeding small'.

The second sense in which these mills came to represent power and energy was due to the scale of the very natural forces which Descartes supposed that they had completely mastered. This proved to be no more exact an estimate of risk than those encountered today in the technology of nuclear power. Exceptional 'natural forces', still feared today and included in insurance documents as 'acts of God', could make these powered wheels highly dangerous. In times of heavy rain the mill-race became a legendary killer. Even the humble windmill that ground the corn so sedately in times of moderate breeze could be a veritable demon during stormy conditions. Records from the early years of windmills showed that many had short lives as they battled with gales. Even the more solid ones built in the eighteenth century, risked fire as well as destruction when strong winds blew.

> 'I was just beginning to uncloth (the sail) when the gale come on like mad and away went the sail as if there was no brake on at all

'...I knew that if the brake were kept on she would catch fire, so I let her off and round she went, the corn flew over the top and smoke blinded me...I put the brake on once again, sparks were flying out all around as she groaned and creaked with strain, but still did not stop.' (Quoted in Waghorn, 1985)

This was a far cry from the musical model which serves to represent this past technology to the infants of today.

The defining technology of power from the Middle Ages remained potent for long enough to serve as an analogy for the first theoretical scientist of the succeeding age of steam power. In 1824 when Sadi Carnot wrote the first mathematical analysis of the steam engine, he used the image of the waterwheel, with heat (caloric fluid) flowing down a temperature gradient, just as water flowed down a gravitational gradient, to produce useful work. The analogy was not perfect since heat is not conserved, as a substance like water must be, but this acknowledgement of a dominant technology from the past was a sure sign of its continued hold on the cultural imagination.

Perpetual Motion

The greatest tribute to any technology must be that it had impetus enough to create an obsession which could tease the imaginations of inventors for several centuries. That is just what the technology of powered wheels did by stimulating the search for perpetual motion. It is true that this fruitless endeavour had valuable spin-offs, not unlike the search for the alchemist's stone which generated a lot of practical chemistry. The most important spin-off was a final theoretical consensus about its total impossibility. However, there were also underlying factors in the search itself which are worth some investigation in order to understand the uneasy interregnum in power technology between the perfect motion of the heavens and the use of energy on earth.

Powered wheels for mills operated without either a theory or a measurable concept of energy. Good mills ground a lot of corn; poor ones ground little or broke down. While engineers learnt to make cogs, gears and cranks they still had no substantial idea about efficiency since they could measure neither the input nor the output of energy. One point, however, was clear — there was always wasted energy.

Of course, friction is a fact of life to engineers in any age. Everything from the wheels of the farm cart to the turbulence in the mill pond, was brought to rest by friction. But it was also true that friction with air or water was the power linkage which drove their machinery. The water leaving the mill still had movement, and the air round the outside of the sails of a rotating windmill seemed more turbulent than average. Here was wasted power, apparently generated by a turning wheel, which might be fed back again into the system to keep it going. Some may imagine that 'feedback' is the invention of modern cybernetics: in reality it was the guiding principle which lured all the inventors of perpetual machines to their fruitless obsession.

In later ages perpetual motion machines became little more than gimmicks and tricks to deceive patrons or public into parting with more money or enhancing a failing reputation. In earlier times, however, the connection between the technology of powered wheels and the designs for perpetual motion by honest and hopeful men were clear enough. The two figures opposite illustrate perpetual motion devices based on the waterwheel and the windmill respectively. Both attempt to feed spent power back into the mechanism to keep it going. Understanding neither the relation between work and energy, nor its transference between systems, the inventors of these projects were not, perhaps, quite as wrong-headed as they sound. At the very least they demonstrate another debt to a dominating technological idea.

At the same time there was a breed of men who had a gut reaction that the whole idea of perpetual motion must be absurd. Leonardo Da Vinci, for example, wrote:

> O speculators about perpetual motion, how many vain chimeras have you created in the like quest? Go and take your place with the seekers after gold. (Quoted from *Notebooks of Leonardo da Vinci*, MacCurdy, 1941)

How could men like Leonardo be so sure? They knew no more physics than their contemporaries, the concept of energy had not been named or defined, and certainly no one had put the Conservation Principle into words. Nevertheless Leonardo and others had this certainty that perpetual motion could not happen; energy just could not pour out endlessly from continuously rotating wheels. Some expressed this in the Latin tag '*Ex nihil nil fit*' which translates most smoothly into Yorkshire dialect 'You can't get owt for nowt.'

But it is not enough to have a piece of common wisdom which, like other mottos and proverbs, can be relied upon in some circumstances but not in others. The only sure test of someone's belief that a principle is universal and unchanging is their willingness to use it as a platform on which to stand another argument. So it was for some early scientists that the impossibility of perpetual motion was more than just obvious, it became itself a physical principle.

Simon Stevinus was a mathematician who lived most of his life in Bruges (1548–1620). In a famous proof he set out to show that the four equal weights (P+Q+R+D) could be in equilibrium with the two similar weights (E+F) if they hung in the directions given by the inclinations of the

The Defining Technologies of Energy

Perpetual motion through feedback

Zimara's perpetual bellows (Ord-Hume, 1977)

Robert Fludd's closed cycle mill (Ord-Hume, 1977)

Getting to Know about Energy in School and Society

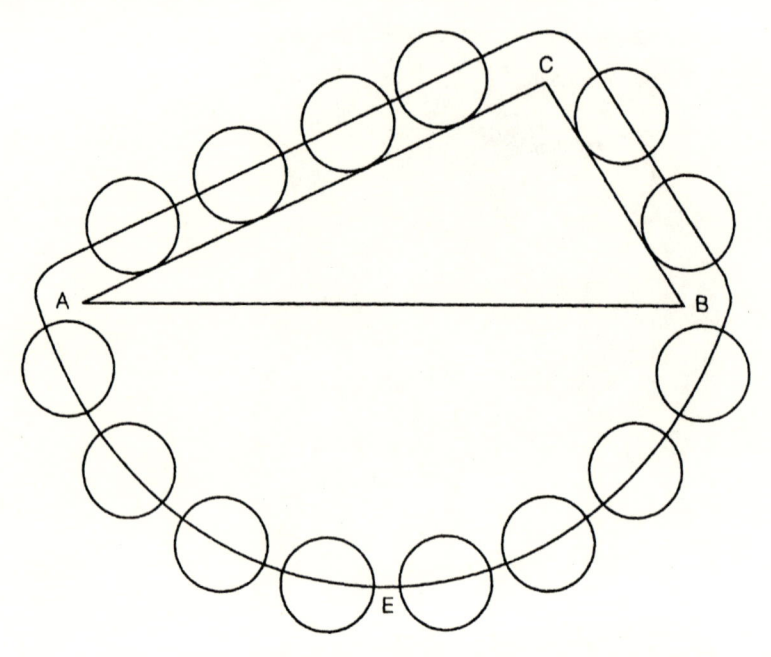

Stevinus showed that fourteen uniform balls on a uniform chain arranged on a triangle ABC so that four balls lay along AC and two along CB were balanced by the eight balls on the curve AEB.

lines AB and BC respectively. The punch line of his proof is that if they were not in equilibrium then the loop of weights would keep moving in the direction of the heavier weight. 'This motion would have no end', wrote Stevinus, 'which is absurd'.

Nearly three centuries later, in 1839, when the Principle of Conservation of Energy had still not been formulated, Michael Faraday showed the same strong intuitive belief in it that Stevinus had. At this time he was having a correspondence with Volta about where the electric power in his new battery came from. Volta favoured the contact between different metals (copper and zinc) as the source of the electric current. Faraday thought there might be some chemical origin to the effect. At any rate he was prepared to argue that Volta's explanation about power coming from a contact, which could be maintained as long as maybe, must be wrong. He based his refutation squarely upon this unspoken physical principle.

> By the great argument that no power can ever be evolved without the consumption of an equal amount of the same or some other power there is no creation of power; but contact would be such creation. (Pearce Williams, 1965:367)

Once the Conservation of Energy became generally accepted all idea about perpetual motion had to stop. But by that time, as we shall see, the contemporary controversy was about another energy notion, and another technology.

Horsepower and Harness

About the same time as windmills were beginning their march across Europe, horsepower was also being discovered. This is all the more surprising since horses of many breeds, including some close to the famous shire horses which were to become a symbol of agricultural power, already existed in late Roman times. However, it was not until the end of the ninth century that horses began to be shod and harnessed. Our King Alfred the Great noted with great surprise, in his day-book, that in northern France some horses were used for ploughing. William the Conqueror's clerks, who so laboriously listed the wealth of their master's new realm in the Domesday book, wrote always of the 'eight-oxen plough', and never of the horse. Perhaps they used this for assessing land tax on the king's new Saxon subjects.

The technology required for the horse harness called for a major rethink of the way animals moved and breathed. When the ox yoke was used for horses it half throttled them; the harder they pulled the less they could breathe. Once a successful horse-collar had been designed the big horses which had been bred to carry soldiers in heavy armour, could begin to apply their strength to the plough, the loaded cart, and the pump. Trucks were drawn by horses along rails before the steam engine replaced them. Technical words to describe working horses became deeply rooted in the language, as references to substantial and continuous power. Even after half a century of the tractor the words are still in very common use. Internal combustion cars boast their 'brake horsepower' and nuclear power is 'harnessed' to the national grid!

There is no doubt that horsepower became a defining technology, in Bolter's sense, and lasted over very many centuries. In one special circumstance it taught the labourers of England more about power than did the wheels that natural forces turned for them. The animals — oxen or horses — had to be fed. The ox ate less and did less work. There is even some extant discussion (*see* White, 1962:62), dating back to the thirteenth century, debating whether the ox or the horse was the more advantageous. The input was hay or other fodder. The output was labour at the plough or treadmill. In a rudimentary sense the notion of energy efficiency was at last taking root.

By the beginning of the nineteenth century ideas of cost efficiency began to rule the boardrooms of industry. The French engineer Lazare Carnot wrote that one horse was equal in energy to about seven men. 'A horse by himself will raise as much water from the bottom of a well of given depth as seven men working together in the same time.' Then Thomas Young defined energy more generally, as 'the tendency of a body...to penetrate to a certain distance in opposition to a retarding force' — much as our pupils learn it today. He was interested in the economics of mechanical power and comparative running costs. 'The expense of keeping a horse', we are coolly informed, 'is twice or thrice as great as the hire of a day labourer.' The horse's extra capacity for work made their energy 'about half as expensive as that of men.' His calculation for the cost efficiency of steam engines presaged the death-knell of working horses.

> 84lb of coal (for the steam engine for a day) is equivalent to the daily labour of 8⅓ men, or perhaps more...the expense of the machinery generally renders a steam engine somewhat more than half as expensive as the number of horses for which it is substituted.

Last, and always least, was the donkey. This came to mind recently, as I was digging in the garden and came across an iron shoe so small I was at a loss to know to what animal it could have belonged. My older village neighbours assured me that such donkey shoes used to turn up by the sackload. Donkeys were the poor man's horses and also a humble ancillary symbol of power. When the original steam engineers made a valve which used a little of the power to control the flow of steam, it was called a donkey valve. The small steam and petrol engines that ran on working boats were also commonly named after those humble working animals now almost forgotten as a source of power. Only phrases like 'donkey work' retain a rather precarious place among our common figures of speech.

Flowing Heat

Technology is never static. In the eighteenth century wind, water and harnessed animals continued to 'turn the wheels of industry', but the locality of work was changing and the products were now more varied than ever before. Blast furnaces made iron from coke, glass works were far busier making new crystal ware, and ceramics were decorated with glazes in a vast new range of colours. Fire, and eventually steam, became the masters of the age.

This suggests that it might have been the emerging chemical technologies that would define fire and heat in some new way to contribute another strand to the cultural imagination, yet it hardly figures in the linguistic legacy of energy. Later in the century, and all through the next, coal was king, yet we do not speak of 'mining for power', or 'coal energy' with any fluency. There may be at least two reasons for this. In the first place power and industry had ceased to be a village affair. The workplace became the factory and not the home where comfortable clichés and maxims are most commonly fashioned. It should also be remembered that 'coal-fired' industry (the only related phrase extant) in Ironbridge, like the alkaline industry of the next century, produced life-threatening pollution far worse than any we would tolerate today. There was little, perhaps, which could be recalled with affection about the contemporary energy technology.

The second reason for an absence of impact from the new industrial technologies was uncertainty about the connection between fuel and energy. Coal had been used, on and off, since the eleventh century for burning on the hearth. It was not as good as wood; Elizabeth I hated it for the smell and smoke it made; James I encouraged its use by personal example in order to conserve the country's dwindling woodlands. Fuels were for making heat — but that did not seem to be the same as energy.

Nevertheless, finding a new theme in energy is not hard at all if we look to contemporary science. It was, above all, an age of theoretical fluids. Despite Newton's celestial mechanics both he and others found the existence of gravitational forces within the emptiness of space almost inconceivable. As a young man Newton held fairly fast, for most of the time, to his maxim that he would make no hypotheses about imponderable immeasurable things, but as an older man he succumbed completely. He wrote about 'subtle elastic fluids' in space, and other contemporary scientists were very comfortable with the idea. In the eighteenth century Euler built up a mathematics of fluids using the calculus, and Lavoisier included imponderable fluids such as heat and light in his lists of chemical elements. The problem for us lies not in finding other examples of such weightless fluids — they were thought to be involved in burning, in the carrying of disease, in electricity, in gravity, in magnetism, and most of all in heat — but in trying to understand how they were imagined by either the scientists or common people of the time.

The term 'fluid' included gas and liquid, anything that flows, and they were often also thought to be made up from particles. Scientists used their new and improved balances in an attempt to weigh the caloric or electric fluids, and were quite unperturbed when they got a zero answer. Eighteenth-century writing about these fluids was not short on adjectives; they were almost always 'invisible and mysterious', usually 'elastic and self-repelling', they were 'subtle and spirituous' and thus without viscosity (although occasionally making unseen, but powerful, vortices), often they were 'weightless' and occasionally, as in the case of phlogiston, could even have negative weight. In short, these 'aerioform fluids' were perfect examples of a scientific way of thinking which is incommensurate with our own, to use Kuhn's famous term. For scientists there was nothing theoretically fancy about these fluids. Even that stolid Manchurian schoolmaster, John Dalton, drew his atoms with 'self-repelling atmospheres of caloric fluid' around each one.

So the legacy of the times was a notion that heat 'flowed', and this is with us still. The rest of the paraphernalia of subtle fluids has been taken up in para-science circles where effluvia and weightless ectoplasm are used to describe psychic power. At one point this odd way of thinking may have had a momentary influence on a new field of physics. William Crookes was the first to describe the luminous effects of cathode rays which J.J. Thompson later named electrons. In the 1870s Crookes also showed that, although apparently weightless, the rays could make a thin screen of glass recoil and were deflected by a magnet. He was lyrical in his description of the strange new rays, 'the fourth state of matter' and a 'corpuscular form of light'. But then William Crookes was an ardent spiritualist and later president of the Society for Psychical Research.

In the eighteenth century Joseph Black's work on this heat fluid — caloric — did make some contribution to ideas about energy. Once he had quantified it by measuring 'heat capacities', a term which nicely expresses the filling up of a substance with fluid, like beer into a barrel, the caloric became so real to him that it was essentially conserved. In fact, Black had clearly started with the idea of conservation, as would anyone who supposed a material basis for heat, even if it was weightless. The quantitative experiments he performed served only to strengthen his conviction. When he realized that beating out a soft metal like lead generated heat, he had no difficulty at all in imagining that the caloric fluid was being squeezed out of the interstices of the metal.

The only real problem with the caloric theory was that it was too confined to heat. The next character on the scientific scene was a brilliant and showy ex-American Commander-in-Chief of the Bavarian army. In 1799 Count Rumford performed a classic experiment using horses — appropriate symbols of power — to generate heat by boring out a cannon. The details of this very public experiment are not important, but the argument was interesting. Rumford claimed that because he could produce 'unlimited' amounts of heat, it could not be a conserved fluid as Black had supposed. Paradoxically it was only by taking a step away from conservation in this fashion, and emphasizing the enormous amounts of heat generated with only a few metal chippings to show for it, that Rumford could topple the caloric theory. Then the way was clear for a new dynamical theory of heat and a new associated concept of conservation.

Vitalism and Vitality

Perhaps this rather episodic history of the technologies which have dominated and shaped our thinking about energy should have begun with human energy. As recounted in chapter 1 the *'vis viva'* (living force) of things was used to define energy quantitatively in the time of Leibniz but it must have influenced thinking since time immemorial. Something new happened in the nineteenth century.

At that time biology considered itself to be a poor relation of physics. Only in the discovery of the circulation of the blood had Harvey, consciously using the mechanical pump as an analogy for the heart, made the kind of advance which seemed worthy of Newton's age of mathematical mechanics. In the eighteenth century Stephen Hales had furthered this mechanical analogy by measuring blood pressure and the volume of fluid in the heart. He even used the same mathematics for this as for calculating the volume of drinking water which needed to be carried by pipes to his parishioners at Richmond. This was the sort of mathematical theory that contemporary scientists, biologists included, most admired.

Physics was the dominant paradigm for biologists, but mechanics served the same technological purpose for almost everyone else. Clockwork and steam power had seized the popular imagination to an extent which can be clearly seen in the Victorian books of inventions such as those by de Vries. Even living creatures were mimicked by mechanisms. For many years hurdy-gurdies were a common sight in the streets of northern Europe, and in the collections of the wealthy there were ingenious almost life-sized figures of musicians realistically playing any or all of the instruments of the orchestra. It was the age of the ballet *Coppelia* and of the story of *Pinnochio*.

In the 1840s a group of young scientists working at the university of Berlin found themselves with a common and revolutionary purpose — they aimed to reduce all the problems of physiology to those of simple mechanics. Not for them the elusive vital force which operated living creatures in special ways that made them seem totally different from non-living matter. Living creatures were to obey the same laws of physics, mechanics, and chemistry as everything else; what Newton had done by extending scientific concepts from the earth to the heavens, Helmholtz and his friends aimed to do for living organisms. It was a considerable challenge. The accepted vitalist theory of the times against which they intended to do battle, was quite explicit:

> In living nature the elements appear to obey quite different laws than in the dead state.... When regarded as the object of a chemical examination a living body is a workshop in which numerous chemical processes take place, whose final result is the creation of phenomena the totality of which we call life.... After a certain slow-down in the processes they finally cease, and from that moment the elements of the previously living body begin to obey the laws of unorganic nature. (Berzelius, quoted in Elkana, 1974:103)

Helmholtz was the first to publish a scientific paper in which the Law of Conservation of Energy, in all its manifestations, was clearly stated. What Leonardo and Faraday had felt intuitively to be true, Helmholtz put into unequivocal words and equations. He included fuels and foods as well as kinetic, potential, and electrical energy.

Fortunately much less guesswork is needed to find a current technology to whose metaphorical influence we can attribute this advance, than there

181

was for the fluid theories of the previous century. Helmholtz himself made it abundantly clear. Like several other scientists of the nineteenth century he was keen to pass on his interests and achievements in science by means of popular lectures. This was the period when scientific and literary societies were beginning to flourish (*see* Layton *et al.*, 1986) and Helmholtz gave an illustrated and eloquent account of his conservation principle some nine years later in Königsberg. He began his public lecture as follows:

> Among the practical arts which owe their progress to the natural sciences, from the conclusion of the Middle Ages downwards, practical mechanics, aided by the mathematical science which bears the same name, was one of the most prominent.... The marvel of the last century was Vaucanson's duck which fed and digested its food.... The writing boy of Droz was publically exhibited in Germany some years ago. When, however, we are informed that this boy and its constructor, being suspected of the black art, lay for a time in the Spanish Inquisition, and with difficulty obtained their freedom, we may infer that in those days even such a toy appeared great enough to excite doubts as to its natural origin. (Helmholtz, 1854)

He went on from this description of wind-up simulacra to a review of the doomed search for *perpetual motion*, and linked that with the movement of real living creatures.

> ...Beasts and human beings seemed to correspond to the idea of such an apparatus, for they moved themselves energetically and incessantly as long as they lived and were never wound up; nobody set them in motion. A connexion between the supply of nourishment and the development of force (energy) did not make itself apparent. The nourishment seemed only necessary to grease, as it were, the wheelwork of the the animal machinery, to replace what was used up, and to renew the old. The development of force (energy) out of itself seemed to be the essential peculiarity, the real quintessence of organic life. (Helmholtz, op. cit.)

Helmholtz knew just what he was doing, in rhetorical terms, when he likened the vitalist theory to the discredited belief in perpetual motion. In an industrial age few would accept that an endless amount of work could be obtained without the consumption of some resource. It was not sound business sense. 'Work', said Helmholtz, 'is money.' It must have sounded very convincing to a commercial audience.

The gulf between physics and biology, and their respective status, meant that advances in understanding energy now followed two distinct paths. The physicists' programme was to write mathematical equations to

The Defining Technologies of Energy

describe the working of idealized heat engines working with idealized gases; the biologists pursued their ambition to understand the processes of life. The latter was a more demanding task because it was not at all clear how useful idealization could take place. Biology uses formal symbolic knowledge every bit as much as physics does, but appropriate analogies on which to erect the symbolism may be harder to find.

Helmholtz' analogy to the clockwork toy figure, where chemical energy in food took the place of potential energy in a wound up mechanism, was only satisfactory for one aspect of life energy. Total input and output of energy matched well enough. But living organisms differed from engines in having the power to grow in complexity; while the energy 'fed' to engines was just dissipated, and the machinery deteriorated. At the same time as physicists were constructing alternative statements of the Second Law of Thermodynamics (Running down, *see* chapter 7) biologists were still arguing about the vital force — and for a good reason. Living creatures did not run down, at least not until they became old or ill; during most of life they actually built themselves up using simple materials to make more complex ones.

A famous scientific paper of 1866, nearly twenty years after the publication of the Conservation Principle, comes to the defence of the vital spirit by quoting a common maxim.

> That the vital spirit is *not* a physical power which performs work... is evident from the fact that it may be infinitely extended — *from a single acorn a whole forest of oaks may result*. (Henry, quoted in Elkana, op.cit.:102)

Recourse to everyday proverbs and common-sense knowledge is a sure sign that the analogy is not acceptable. Understanding had to await the arrival of some new technology to produce a better way of thinking about the energy processes of life.

Organization and Information

Growth does occur outside the domain of living organisms: for example crystals grow if left in a slowly evaporating or cooling solution. Berzelius himself identified the difference between these two kinds of growth as one of *organization*. Crystals reproduce themselves uniformly, but living creatures are able to use energy to synthesize the wide variety of different substances which they need, when and where they need them — and their needs are often very complex. They could also reproduce their kind, and through the operation of evolution it seemed to some Victorian biologists that they could even produce changed and 'better' kinds. This was entirely different from crystal growth. The words which seemed to distinguish the

183

Getting to Know about Energy in School and Society

living from the non-living were now 'choice', 'complexity', 'information', 'synthesis'. The problem was not the amounts of energy supplied: Helmholtz's formulation of the Conservation Principle had produced a gross overall balance-sheet for energy which seemed satisfactory. The remaining dilemma was how the energy was controlled and used. During most of the nineteenth century there was no familiar technology to think about questions like that.

The twentieth century had hardly begun before the need to communicate information began spawning one new technology after another. Telegraph and radio had been the first, and by the 1940s and 1950s they had been operating for long enough for some new meanings to have crept into the language and the imagination. There was 'code' to describe the transformation of information which became the 'message', a 'signal' which was to be selected by 'tuning', and 'interference' which spoilt 'reception'. (A sure sign of new meanings was the school child's new burden of ambiguity — 'volume' now meant a knob to increase the sound as well as three dimensional size, and this misconception was soon identified by educational research.) The new information technologies consumed extraordinarily little power compared with the fire and steam of the previous technology. What they did was to receive, decode and select information. Later they also began to use this information to control much larger energy usage.

These technologies were in the domain of physics rather than biology so it was some time before the information theory of the 1950s, which grew out of them, was seen to have relevance to the problems of biology. It measured information in bits and bytes, and calculated its loss of definition. Its formalisms showed a mathematical equivalence between the decay of information and the running down in an energy transformation. Energy input in a radio was required to prevent the loss of signal clarity; this was not just for simple amplification, but for selection of information through more subtle tuning to the correct channel. The public's understanding of electricity is notoriously poor, but once the 'wireless' was installed in the family living room, and later when the 'tranny' became the teenager's constant companion, understanding of how to speak about and operate the technology, although not its circuits, was everywhere available. It created words to inform a new outlook on energy.

The transition from the tuning of a radio in order to hear 'The News', to the selecting and building of materials by an enzyme in a cell, is not simple. But the twentieth century had acquired new maxims about power from the political arena, particularly from Marxism, to provide help with this difficult technological analogy.

> The activity of the cellular enzymes suggests that they are... trading information (knowledge) for negentropy (power to prevent running down), thus demonstrating anew that knowledge is power

The Defining Technologies of Energy

as much within the molecule of life as it is without in the life of man. (Singh, 1966)

Artificial intelligence, with its analogy of brain power to electric circuits, became a practical project once the techniques of electronic 'control' and 'logic' were established. One of the defining technologies of the twentieth century promoted a general understandng of how a small energy input to the living cell could be thought of as controlling and informing the selection of materials for building up the substances of the cells, our own or those of our offspring, to a preselected 'program'. A similar process was at work in the increasingly automated production lines of industry.

It is true that the combined efforts of Franklin, Crick and Watson made the specific and important discovery of the structure of DNA without direct reference to information. But that advance had followed and built upon a bedrock of essential understandings about transfer of information. Almost immediately after the details of the structure were published phrases such as 'the genetic *code*', or '*messenger* RNA' slipped into educated conversation. Without cultural preparation through half a century of information/energy equipment to provide images and meanings from our dominant technology, it is doubtful if the now potent symbol of the double helix would have meant any more to us than, say, the twined serpents of Aesculepios.

Technological Future

Some reflection on the philosophy of technology is necessary in order to make sense of the present, and even future, understanding of energy. New human concerns generate new technologies which need both an energy supply and a way to think about the nature of energy. Out of this we acquire not just new terms and metaphors of the kind which come in handy for constructing scientific theories, or for decorating literary prose. New and pervasive technologies alter ways of thinking at a more profound level.

It has been easy enough to identify a few of the ways of thinking about energy in the past, and to describe them in terms of different human reactions. Energy technology may have stimulated the utopian response of the early seekers after perpetual motion; many centuries later mechanical energy became part of the profit and loss account sheets of Victorian entrepreneurs. Yet it would be absurd to attribute the characteristics of an age, utopianism or commercialism, simply to the arrival of a new technology. Any reflection on the rise and fall of past technologies faces problems of historical causation in special ways. For energy technology these would have to include available materials and social purposes, as well as skills and prevailing ideologies. Each of these will have an important bearing on technological uptake; and yet to separate each 'cause', in an Aristotelian fashion, not only complicates the question enormously, it also obscures any more general perspective upon the character and technology of the times. It would seem to be a more tenable position to assume that the temper of an age and its energy technologies complement each other because they are interactive at many levels. The uptake and exploitation of a technology which is on the point of emerging is both directed by, and also reinforces, the emerging preoccupations of the age.

The new philosophy of technology has spent an inordinate proportion of its short life making sharp distinctions between past and present technology in terms of naturally available materials, intuitive knowledge, locality, and cultural separatism. The standpoint of 'Turing's Man' suggests that such a division is misconceived and the product of a time-restricted vision. Plastic materials, for example, are now as familiar as wood, perhaps even more so, and would be readily used in the kind of handicraft which such philosophers often categorize as an ancient and more desirable technology. The whole conception of new man-made materials which have been specially designed for a job, is now sedimented into the stock of common intuitive knowledge. The same sort of argument can be given to show that the conception of what is local has also changed. At one point of technological change the peasant stopped grinding his own corn and handed the task over to the village miller. Now flour making is a national and international business; but its technology is only a little more arcane to the ordinary citizen than was that of the miller. Technology, past or present, has always demanded specialized knowledge; the main difference is only that the range of knowledge has gradually extended.

To identify how our own age may be altering its thinking about energy in response to technological change requires an understanding of the range of knowledge but also, and more importantly, a re-identification of public interest. We shall need a perception of the place of technology within history, such as that of Bolter's defining technologies, and to extend rather than replace it. Such a point of view focuses upon the integration of technology within a culture and uses general lessons from the past, as well as instances from the present. Complete and uniform integration of technology is a chimera: no culture could be uniform in respect of all its thinkers. However, this emphasis on the internalization of technology within a culture is essential for understanding our present and future predicaments.

> The deepest philosophical analysis of technology attempts to reach that point at which technology takes on its greatest density — the point at which objects, processes, knowledge and volition meet. (Mitcham, 1970:322)

Applying that search for density or technological definition suggests that, in our century, space travel, industrial pollution, and the science of ecology may have come together in a significant way. The 1960s were the first decade of manned space flight. There had been an enormous imaginative investment in the moment when a human being would first leave earth. For more than a century writers had teased their imaginations with stories of how it might be in space, or in other worlds. Like all appetite for vicarious experience it was satiated by the reality, and so there was a corresponding fall in sales of science fiction books during the decade of moon landings. None of the wild predictions of SF matched the immediate reaction of almost every astronaut who was launched into space: one after another they looked back at their home planet and called it 'beautiful'. The earth had never been seen like that before.

Down below it seemed less and less beautiful by the year. The human population doubled and redoubled; industrial pollution spread from the developed world to the Third World. Land-locked lakes were the first to become lifeless, but the smaller seas and marginal over-farmed lands were following fast on their heels. Conservation became a catchword on everyone's lips while the economists and scientists struggled to conceptualize a threatening reality.

The science of ecology had begun slowly in the early years of the century. At first it concerned management of the environment and the 'economy' of nature (the root from which the word ecology had been coined by German nineteenth-century Darwinists). It also resembled economics in several ways as might have been predicted by our technological thesis. Commenting on the use of phrases such as 'assembly line conveying energy', and 'the efficiency of a river's productivity', Wooster wrote:

> The metaphors used here are more than casual or incidental; they express the dominant tendency in the scientific ecology of our times. In their most recent theoretical model ecologists have transformed nature into a reflection of the modern corporate state. (Wooster, 1977:292)

In the hands of the Oxford botanist A.G. Tansley, ecology blossomed into a typical science using established concepts and theories, including those of energy. 'Energy flow' through 'a balanced ecosystem', partially replaced earlier terms such as 'food-chain' and 'community'. Discussion of the Second Law of Thermodynamics, 'energy budget' and 'energy efficiency' placed this twentieth century scientific discipline firmly in the energy tradition.

However, it was not just protected areas of woodland which were to be the ecologists' laboratory; whole countries, and then the planetary globe itself, became involved. Sewage in the seas washed up on several shores, and the excess of carbon dioxide in the atmosphere from the burning of fossil fuels was distributed worldwide. Suddenly the scientific discipline became almost overwhelmed by public and political attention. People whose health was at risk, and whose education was now better than ever before, wanted and needed to understand. They read avidly, the more sensational the literature the easier they found it to absorb and so, in the manner of informal learning (chapter 5), some of the lessons of ecology were added to their life-world stock of knowledge about energy and their emotional reactions to its generation and use.

Energy from burning fuels is not only a resource in short supply, it has now become an ecological threat which neither politicians nor scientists know how to understand. The Green Party writes that 'we are slowly killing our planet because of (cars)' (Kemp and Wall, 1990). The new science of Gaia attempts to see our planet as a self-regulating organism. Beneath both these approaches is a severe difficulty in understanding which is quite unrelated to technological, scientific, or economic matters, as they have previously been conceived. The difficulty lies in comprehending our moral obligations (Taylor, 1986). Where our well-being is at risk, self-interest prompts us to apply a remedy. If our children, or grandchildren might be impoverished we may still feel some obligation to act. But thinking about how we should use power for industry and yet protect wildlife, or heat our homes and yet ensure the planet's future, is altogether more confusing. Never before has learning about energy seemed to involve such considerations, for it is only in our age that energy conservation, in the non-physics sense, has seemed set to become a defining technology.

Ecological morality is an entirely unexpected addition to the austere study of thermodynamics. Perhaps the next step in getting to know about energy will be learning, arguing and discussing the ethics of earth management.

References

APU (1983, 1984) Assessment of Performance Unit, Science, HMSO London.
ARMSTRONG, M. (1980) *Closely observed children: The diary of a primary classroom*, Writers and Readers, London.
ASHE, S. (1956) 'Studies of independence and conformity', *Psychological monographs*, **70**(9).
BAKER, K. (1989) 'Science and the National Curriculum in England and Wales, *Physics Education*, **24**(3), p. 117.
BALFOUR STEWART, L. (1870) 'What is energy?', *Nature*, 28 April, p. 647.
BARNES, D. (1976) *From communication to curriculum*, Penguin, Harmondsworth.
BARNES, D. and TODD, F. (1977) *Communication in small groups*, Routledge and Keegan Paul, Henley.
BIJKER, W., HUGHES, T. and PINCH, T. (Eds) (1987) *The social construction of technological systems*, MIT Press, Mass.
BILLIG, M. (1987) *Arguing and thinking*, CUP, Cambridge.
BLACK, D. and SOLOMON, J. (1987) 'Can pupils use taught analogies for electric current?', *School Science Review*, **69**(247), pp. 249–54.
BLACK, M. (1962) *Studies in language and philosophy*, Cornell University Press, Ithaca.
BLUMER, J., NOSSITER, J. and McQUAIL, D. (1976) *Political information and the young voter*, SSRC Report.
BODMER, W. (1985) *The public understanding of science*, The Royal Society, London.
BOLINGER, D. (1975) *Aspects of language*, Harcourt, Brace, Jovanovitch, New York.
BOLTER, D. (1987) *Turing's man*, Penguin, Harmondsworth.
BOURDIEU, P. (1977) *Outline of a theory of practice*, Cambridge University Press.
BREAKWELL, G. (1990) *Young people's attitudes to scientific change*, paper presented at the Policies and Publics for Science and Technology Conference, The Science Museum, London.
BRIDGES, D. (1979) *Education, democracy and discussion*, NFER, Windsor.
BRONOWSKI, J. (1964) *The common-sense of science*, Penguin, Harmondsworth.
BROOK, A. and DRIVER, R. (1986) *The construction of meaning and conceptual change in classroom settings: case studies of energy*, University of Leeds.
BROWN, J. (1976) *Recall and recognition: Especially ecphoric processes*, E Tulving, Wiley.
BRUNER, J. (1978) *Towards a theory of instruction*, Belknapp, Harvard.
CHAMPAYNE, A., KLOPFER, L., SOLOMON, C. and CAHN, A. (1980) *Interaction of students' knowledge with their comprehension and design of experiments*, University of Pittsberg.

189

References

CLAXTON, G. (1984) 'Teaching and acquiring scientific knowledge', in KEEN, T. and POPE, M. (Eds) *Kelly in the Classroom*, Cybersystems, Montreal.
CLIS. WIGHTMAN, T., GREEN, P. and SCOTT, P. (1986) *The construction of meaning and conceptual change in the classroom*, University of Leeds.
COLLINGS, P. and SMITHERS, A. (1984) 'Person orientation and science choice', *International Journal of Science Education*, 6(1), pp. 55–65.
COLLINS, P. and BODMER, W. (1986) 'The public understanding of science', *Studies in Science Education*, 13, pp. 96–104.
DESCARTES, R. (1637) *Discourse on method.*
DISESSA, A. (1981) *Phenomenology and the evolution of ideas*, DSRE working paper, MIT Cambridge.
DOISE, W. and MUGNY, G. (1984) *The social development of the intellect*, Pergamon Press, Oxford.
DONALDSON, M. (1978) *Children's minds*, Collins, Glasgow.
DRIVER, R. (1983) *The pupil as scientist?* Open University Press, Milton Keynes.
DRIVER, R. and EASLEY, J. (1978) 'Pupils and paradigms', *Studies in Science Education*, 5, pp. 61–84.
DUIT, R. (1981) 'Understanding energy as a conserved quantity', *European Journal of Science Education* 3(3), pp. 291–301.
DUIT, R. and TALISAYON, V. (1981) *Comprehension of the energy concept: Philippine and German experience*, Conference on Energy Education, Providence RI.
DURANT, J., EVANS, G., THOMAS, G. (1989) 'The public understanding of science', *Nature*, 340, pp. 11–14.
EDGE, D. (1985) 'Dominant scientific methodological views: Alternatives and their implications', in GOSLING and MUSSCHENGA (Eds) *Science Education and Ethical Values* Georgetown University Press, Washington.
EIJKELHOF, H. (1990) *Radiation and risk in physics education*, Centre for Science and Mathematics Education, University of Utrecht.
EIJKELHOF, H. and MILLAR, R. (1988) 'Reading about Chernobyl: The public understanding of radiation and radioactivity', *School Science Review*, 70(251), pp. 35–41.
ELKANA, Y. (1974) *The discovery of the conservation of energy*, Hutchinson Educational, London.
ELKANA, Y. (1983) *The borrowing of the concept of energy in Freudian psychoanalysis*, Leo Olschki, Florence.
ELLIOTT, J. (1978) 'What is action research?' *Journal of Curriculum Studies*, 19, pp. 355–7.
ENGELS, E. (1982) 'The development of understanding of selected aspects of pressure, heat and evolution, unpublished Ph.D. thesis, University of Leeds.
ERICKSON, G. (1976) 'Children's conceptions of heat and temperature', *Science Education*, 60, pp. 221–30.
FESTINGER, L. (1962) 'Cognitive dissonance', *Scientific American*, October issue.
FEYERABEND, P. (1978) *Against method*, Verso, London.
FEYNMAN, R. (1963) *Lectures on physics*, California Institute of Technology.
FLANDERS, N. (1970) *Analyzing teacher behaviour*, Addison Wesley.
GALILEO (TRANS, H. Crew 1941) *Dialogues concerning two new sciences*, Dover Macmillan, New York.
GARDNER, H., BECHOFER, K., WINNER, E. and WOLFE, D. (1979) 'Figurative language', in *Children's Language*, Gardner Press, New York.
GILBERT, J., OSBORNE, R. and FENSHAM, P. (1982) 'Children's science and its consequencies for teaching', *Science Education*, 66(4), pp. 623–33.
GILBERT, J. and WATTS, M. (1983) 'Concepts, misconceptions and alternative conceptions: Changing perspectives in science education', *Studies in Science Education*, 10, pp. 61–98.

References

GILLIGAN, C. (1983) *In a different voice*, Harvard University Press.
GOFFMAN, E. (1959) *The presentation of self in everyday life*, Penguin, Harmondsworth.
GUESNE, E. (1976) 'Lumière et vision des objects: un exemple de representations des phonomenes physiques preexistant a l'eseinement', *Proceedings of Girep*, Taylor and Francis, London.
HALLIDAY, M. (1978) *Language as social semiotic*, E. ARNOLD, London.
HEMPEL, C.G. (1952) 'Fundamentals of concept formation in empirical sciences', in *Foundations of the Unity of Science*. (**Vol.** 2) University of Chicago.
HESSE, M. (1966) *Models and analogies in science*, University of Notre Dame Press, Paris.
HARRE, R. (1979) *Social being*, Basil Blackwell, Oxford.
HEAD, J. (1981) *The personal and affective aspects of learning*. Paper given at the Oxford Conference on Science Education.
HEAD, J. (1984) *The personal response to science*, Cambridge Educational Press.
HELMHOLTZ, H. (1962) *Popular Scientific Lectures*, (Ed.) KLINE, M., Dover, New York.
HEWSON, P. (1981) 'A conceptual change approach to learning science', *European Journal of Science Education*, 3, pp. 383–96.
HODGE, B. and TRIPP, D. (1986) *Children and television*, Polity, Cambridge.
HODGKIN, R. (1985) *Playing and exploring. Education through the discovery of order*, Methuen, London.
HUSTLER, D., CASSIDY, T. and CUFF, T. (1986) *Action research in classrooms and schools*, Allen and Unwin, London.
INHELDER, B. and PIAGET, J. (1958) *The growth of logical thinking*, Routledge and Keegan Paul, London.
IRWIN, A., DALE, A. and SMITH, D. (1990) *Science in hell's kitchen*, University of Manchester.
JENKINS, E. (1990) *Domestic energy resources and the elderly: The understanding of energy*, Conference paper, Policies and Publics for Science and Technology, Science Museum, London.
KELLY, G. (1955) *The psychology of personal constructs*, Norton.
KEMPA, R. and HODGSON, G. (1976) 'Levels of concept acquisition and concept maturity in students of chemistry', *British Journal of Educational Psychology*, **46**, pp. 253–60.
KUHN, T. (1962) *The structure of scientific revolutions*, University of Chicago Press.
LARKIN, J., MCDERMOTT, L., SIMON, D. and SIMON, H. (1980) 'Expert and novice performance in solving physics problems', *Science*, **208**, pp. 1335–42.
LAYTON, D. (1984) *The alternative road*, University of Leeds.
LAYTON, D., DAVEY, A. and JENKINS, E. (1986) 'Science for specific social purposes', *Studies in Science Education*, 13, pp. 27–52.
LEBOUTET-BARRELL, L. (1976) 'Concepts of mechanics among young people', *Physics Education*, pp. 462–5.
LOWRANCE, W. (1967) *Of acceptable risk*, Kaufmann, Los Alton Cal.
MAK, SE-YUEN, and YOUNG, K. (1987) 'Misconceptions in the teaching of heat', *School Science Review*, **68**(244), pp. 464–70.
MARSH, P., ROSSER, E. and HARRE, R. (1978) Routledge and Keegan Paul, London.
MARTIN, N., WILLIAMS, P., WILDING, J., HEMMINGS, S. and MEDWAY, P. (1976) *Understanding children talking*, Penguin, Harmondsworth.
MCCLOSKEY, M. (1983) 'Intuitive physics', *Scientific American*, **248**, pp. 122–30.
MCGILL, S. (1987) *The politics of anxiety*, Pion, London.
MEAD, G.H. (1934) *Mind, self and society*, University of Chicago Press.
MERTON, R. (1973) *The sociology of science*, University of Chicago Press.

References

MEYROWITZ, J. (1984) 'The adultlike child and the childlike adult: Socialization in an electronic age', *Deadalus*, **113**(3), pp. 19–48.

MILLAR, R. (1989) 'Constructive criticisms', *International Journal of Science Education*, **11**(5), pp. 576–87.

MILLAR, R. and WYNNE, B. (1988) 'Public understanding of science: From contents to processes', *International Journal of Science Education*, **10**(4), pp. 388–98.

MILLER, J. (1983) 'Scientific literacy: A conceptual and empirical review, *Daedalus*, Spring issue, pp. 29–48.

MITCHAM, C. (1980) 'The philosophy of technology', in DURBIN, P.T. (Ed.), *A Guide to the Culture of Science, Technology and Medicine*, Free Press, Macmillan, London.

MONK, M. (1990) *Genetic epistemological notes on recent notes on children's understanding of light*, Kings College, London.

MORI, I., KOJIMA, M. and TADAING, K. (1976) 'The effect of language on a child's conception of speed: A comparative study on Japanese and Thai children', *Science Education*, **60**(4).

MOSCOVICI, S. (1976) *Social influence and social change*, European monographs on social psychology, London Academic Press.

NEEDHAM, J. (1962) *Science and civilization in China*, CUP, Cambridge.

NIXON, J. (1987) 'Only connect: Thoughts on stylistic interchange within the research community', *British Educational Journal*, **13**(2), pp. 191–202.

ORD-HUME, A. (1977) *Perpetual motion: The history of an obsession*, Allen and Unwin, London.

OSBORNE, J. and FREEMAN, J. (1989) *Teaching physics: A guide to the non-specialist*, CUP, Cambridge.

OSBORNE, R., FREYBERG, P., TASKER, R. and STEAD, K. (1980) *Reconsidering the framework*, University of Waikato, Hamilton.

OSBORNE, R. and WITTROCK, M. (1985) 'The generative learning model and its implications for science education', *Studies in Science Education*, **12**, pp. 59–87.

PEARCE WILLIAMS, L. (1965) *Michael Faraday*, Chapman and Hall, London.

PEPPER, S. (1942) *World hypotheses*, University of California Press, Los Angeles.

PERRET-CLERMONT, A. (1980) *Social interaction and cognitive development in children*, Academic Press, London.

PIAGET, J. (1926) *The language and thought of the child*, Routledge and Keegan Paul, London.

PIAGET, J. (1929) *The child's conception of the world*, Routledge and Keegan Paul, London.

PIAGET, J. (1952) *Play, dreams and imitation in childhood*, Routledge and Keegan Paul, London.

PIAGET, J. (1967) *Biologie et Cognaisance*, Gallimard, Paris.

PIAGET, J. (1972 trans.) *Psychology and epistemology*, Penguin, Harmondsworth.

PINES, L. (1978) 'Scientific concept learning in children', unpublished Ph.D. thesis, Cornell University.

PINES, L. and WEST, L. (1986) 'Conceptual understanding and science learning: An interpretation of research within a sources-of-knowledge framework, *Science Education*, **70**, pp. 503–604.

POLANYI, M. (1958) *Personal knowledge*, Routledge and Keegan Paul, London.

POPE, M. and WATTS, M. (1988) 'Constructivist goggles: Implications for process in teaching and learning physics', *European Journal of Physics*, pp. 101–9.

PREECE, P. (1984) 'Intuitive science: Learned or triggered?' *European Journal of Science Education*, **6**(1), pp. 7–10.

QUALTER, A., STRANGE, J., SWATTON, P. and TAYLOR, R. (1990) *Explorations: A way of learning science*, Blackwell Education, Oxford.

References

RAVETZ, J. (1971) *Scientific knowledge and its social problems*, Oxford University Press.
RAWLS, J. (1971) *A theory of justice*, Harvard University Press.
Ross, K. (1989) 'A cross cultural study of people's understanding of fuels and the process of burning', unpublished Ph.D. thesis, University of Bristol.
Ross, K. and SUTTON, C. (1982) 'Concept profile and the cultural context', *European Journal of Science Education*, 4(3).
ROWELL, J. and DAWSON, C. (1989) 'Towards an integrated theory and practice for science education', *Studies in Science Education*, 16, pp. 47–73.
RUDDUCK, J. (1979) *Learning to teach through discussion*, CARE publication, University of East Anglia.
RUMELHART, D. and NORMAN, D. (1981) 'Analogical processes in learning', in ANDERSON (Ed.) *Cognitive skills and their acquisition*, Lawrence Erlbaum, Hillsdale.
SCHMIDT, G. (1982) 'Energy and its carriers', *Physics Education*, 17, pp. 212–8.
SCHON, D. (1983) *The reflective practitioner*, Temple Smith, London.
SCHUTZ, A. and LUCKMANN, T. (1973) *Structures of the life world*, Heinemann, London.
SHAYER, M. and ADEY, P. (1981) *Towards a science of science teaching*, Heinemann, London.
SHAYER, M. and ADEY, P. (1990) 'Accelerating the development of formal thinking in middle and high school students', *Journal of Research in Science Teaching*, 27(2).
SHEN, B. (1975) Scientific literacy and the public understanding of science, in DAY, S. (Ed.) *Communication of scientific information*, Karger, Basel.
SIMPSON, M. and ARNOLD, B. (1982) 'The inappropriate use of subsumers in biology learning', *European Journal of Science Education*, 4(2).
SINGH, J. (1966) *Great ideas in information theory, language and cybernetics*, Constable, London.
SKILBECK, M. (1984) *School-based curriculum development*, Harper, London.
SMITH, E. and LOTT, G. (1983) 'Teaching for conceptual change. Some ways to go wrong', *Proceedings of the international seminar on Misconceptions in Science and Mathematics*, Cornell University.
SOLOMON, J. (1980) *Teaching children in the laboratory*, Croom Helm, London.
SOLOMON, J. (1982) 'Does the First Law come first?', *Physics Education*, (82), pp. 415–22.
SOLOMON, J. (1985) 'Learning and evaluation: A study of school children's views on the social uses of energy', *Social Studies of Science*, 15, pp. 343–71.
SOLOMON, J. (1985b) 'Classroom discussion: A method of research for teachers?' *British Journal of Educational Research*, 11(2), pp. 153–62.
SOLOMON, J. (1986) 'Children's explanations', *Oxford Review of Education*, 12(1), pp. 41–51.
SOLOMON, J. (1989) 'A study of behaviour in the teaching laboratory', *International Journal of Science Education*, 11(3), pp. 317–26
SOLOMON, J. (1989b) 'A study of behaviour in the teaching laboratory', *International Journal of Science Education*, 11(3), pp. 317–26.
SOLOMON, J., BLACK, P. and STUART, H. (1987) 'The pupils' view of electricity revisited: Social development or cognitive growth?' *International Journal of Science Education*, 9(1), pp. 13–22.
SPENT, P. (1988) *Taking risks: The science of uncertainty*, Penguin, Harmondsworth.
STENHOUSE, L. (1969) 'The nature and interpretation of evidence', in *Authority, education and emancipation* (1983) Heinemann, London.
STENHOUSE, L. (1975) *An introduction to curriculum research and development*, Heinemann, London.

References

STRIKE, K. and POSNER, G. (1982) 'Conceptual change and science teaching', *European Journal of Science Education*, **4**(3), pp. 231–40.

SUMMERS, M. (1983) 'Teaching heat: An analysis of misconceptions', *School Science Review*, **64**(229), pp. 670–6.

SUTTON, C. (1980) 'The learner's prior knowledge: Science language and meaning', *School Science Review*, **62**, pp. 47–56.

TENNEY, Y. and GENTNER, D. (1984) *What makes analogies accessible?*, paper presented at conference on Electricity Education, Ludwigsberg.

TIBERGHIEN, A. and DELACOTE, G. (1976) 'Conceptions de la chaleur chez les enfants 10 à 12 ans', *Proceedings of Girep*, Taylor and Francis, London.

UR, P. (1988) *Discussions that work: Task-centred fluency practice*, CUP, Cambridge.

VAN DER VALK, A., LINJNSE, P., TACONIS, R. and BORMANS, H. (1988) *A research-based strategy for teaching about energy*, University of Utrecht.

VIENNOT, L. (1979) 'Spontaneous reasoning in elementary dynamics', *European Journal of Science Education*, **1**(2).

VYGOTSKY, L.S. (1962) *Thought and language*, MIT Press, Cambridge, MA.

VYGOTSKY, L. S. (1979) *Mind in society*, Harvard University Press, Cambridge, Mass.

WAGHORN, M. (1985) *Brill windmill*, The Brill Society, Oxford.

WARREN, J. (1986) 'At what stage should energy be taught?' *Physics Education*, **21**, pp. 154–5.

WATTS, M. (1983) 'Some alternative views on energy', *Physics Education*, **18**, pp. 213–6.

WEISENMAYER, R., MURRIN, M. and TOMERA, A. (1984) 'Environmental education related to issue awareness', in IOZZI, A. (Ed.) *Monographs in environmental education*, ERIC Clearing House, Ohio State University.

WELLINGTON, J.J. (1986) *Controversial issues in the classroom*, Blackwell, Oxford.

WHITE, L. (1962) *Medieval technology and social change*, Oxford University Press.

WHITEHOUSE, J. (1989) 'How do we improve research-based professionalism in England?', *British Journal of Educational Research*, **15**(1), pp. 3–18.

WHORF, B. (1956) *Language thought and reality*, Massachusetts Institute of Technology Press.

WOOLNOUGH, B. (Ed.) (1991) *Practical science*, Open University Press, Milton Keynes.

WOOSTER, D. (1977) *Nature's economy: A history of ecological ideas*, CUP, Cambridge.

WYNNE, B. (1988) *The sheep farmers and the scientists*, paper given at the Public Understanding of Science Conference, SPSG, University of Lancaster.

WYNNE, B. (1990) 'The Blind and the Blissful', *The Guardian*, 12 April.

YOUNG, T. (1807) *Lectures on natural philosophy* **Vol. 1** (Quoted in WOOD, A. (1954), *Thomas Young, Natural Philosopher*, C.U.P.)

ZEMANSKY, M. (1957) *Heat and thermodynamics*, McGraw Hill, New York.

ZIMAN, J. (1968) *Public knowledge. The social dimension of science*, Cambridge University Press.

ZIMAN, J. (1978) *Reliable knowledge*, Cambridge University Press.

ZIMAN J. (1980) *Teaching and learning about science and society*, CUP.

Index

(Page numbers in *italics* are illustrations)

ability: of children, in energy concept, 57
abstract: definition, in language, 74
  spirit, teaching and learning in, 119–20
  theories, and scientific knowledge, 98, 102–3
academic knowledge: abstract, vi, 98–9
action: potentiality of, 138
action research: in classroom, vi, 130, 136–40
  programmes of, 30
active learning, 37
activity: and energy, 43–4, 72, 74, 94–6
  as life force, 44
adult: culture, 4
  general knowledge, 3–4
    barriers to, 84–5
    from television, 143
    and understanding energy, 83–4
  lack of success, in school science, 85
affective: involvement, 144, 145
  overtones, 68
alternative energy resources, 143, 144, 160
alternative frameworks, 21–2, 24–6, 30
  and constructivism, 37
analogies: children's use of, 105–6
  and scientific theories, 103–5
animal energy, 9
anti-matter, 14
answers: 'correct', 59, 61
Arnold, Brian: and children's misconceptions, 28
artificial intelligence, 36, 185
atomic energy, 6

Barnes, Douglas, 63–4
biology: and information theory, 184
  mechanics of, 180–1
Black, Joseph, 180
Black, Max: and metaphors of energy, 15
body feelings: of exhaustion, 45
  of fitness, 45
boys: responses to energy, 45–7
  see also gender
brainstorming discussion, 65
breath: and energy, 43
Bruner, Jerome, 62

CBA, cost/benefit analysis, 159
CDT energy unit, 89
Champayne, Audrey: and Children's Science, 24
change: cognitive, 29–30
  conceptual, 28–30
Chernobyl incident, 150–1, 153, 156, 163
children: general knowledge of, 3–4
  pre-scientific notions, 31
  social skills of, 80
  see also ideas: children's
Children's Science, 23–5 *passim*
citizen knowledge, 83–4, 141–66
classroom: performance, 61
  talk, 65
cliché: and dead metaphor, 15
CLIS (Children's Learning In Science) Project, Leeds University, 30–1, 38
coal: role in technology, 178–9
cognitive: change, 29–30
  conflict, ix

195

Index

development, 73
dissonance, 29
irritability, of common-sense knowledge, 6
science, 36–7
colloquial speech: and meaning of energy, 7
common sense: cognitively irritable, 6
intersubjective, 6
knowledge, 4–6, 62, 79, 98, 183
and logic, 30
and meaning variation, 53
and physics, 81
security of, 5
taken for granted, 5
thinking, 168
computer: and memory, 169
concepts: abstract, 100
forming, 54–7
links between, viii
conceptual: change, 28–30, 107
understanding, 93–4, 124
consensus: class, 76–7
in scientific community, 149–50
talking towards, 65–9, 75
conservation: of energy, 187, 188
Conservation of Energy Principle, viii, 83, 125, 146, 176, 181–3
intuitive knowledge of, 121–3
rewording, 131
teaching about, 116, 126–40
see also storage
constructive alternativism, 37
constructivism, 37–8, 117
and alternative frameworks, 37
context: of discussions, 59–60
and scientific knowledge, 98, 99–100, 108
controversy: learning through, 153–5
Crookes, William, 180
cultural scientific literacy, 83
culture: meanings in, 51
and perceptions, 32
and technology, 168

Da Vinci, Leonardo: and perpetual motion, 174
Davy, Humphrey: lectures by, 12
debating skills, 154, 165
definition: of energy, and amalgamation, 16
degradation of energy, 116
Delphi method of syllabus design, 149–50

destruction: Kelvin's mathematical scale of, ix
development: cognitive, 73
internal mental stages of, 20
developmental epistemology, 74
discussion sessions, 59, 144, 154
chaining effect, 67–73
inter-pupil, 61, *65*
small group, 155–7
teacher-pupil, 59, *65*
see also brainstorming discussion; DISS; pseudo-discussion; trails
diSessa, Andrea: and primitive notions, 31, 45, 91
DISS (Discussion of Issues in School Science) Project, 155–7, 162
Donaldson, Mary, 79, 83, 98
drawing: and energy concept, 57, *58*
Driver, Rosalind: and participant observer research, 21

earth management: ethics of, 188
ecology: morality of, 188
science of, 187
education: as social reconstruction, 141
technology in, 167–8
effort: and energy, 43
ego-centricity: in children's thinking, 21, 46
and energy, 44–5
electrical energy, 43, 48–9, 82
children's explanations, 24–5
electric current, 38
electricity, 50, 55, 144
Nigerian conception of, 35
and similes, 105
emotional reactions: by children, 39
empathy: and energy knowledge, 161–2
and social justice, 163–5
energetic, 34, 46, 53, 82
energeticness, 53, 72, 76, 81, 93, 122
energy: abstract nature of, 9
ancient view of, 8
budgeting, for elderly, 87
as cause, 72
changes, 131, 132
of character, 13
definition, 16
general knowledge about, 1–17
historical meanings, 7–8
and *Kraft*, 34
measurement of, 4, 14, 54
mental control of, 9
pure, 8, 9

196

Index

symbolism of, 16
transfer, 94, 100, 102, 106-7, 121
transformation, 102, 136
energy crisis: impact of, 142-4, 146
energy laws, viii
  see also Thermodynamics: Laws of
Engels, Liz: and Children's Science, 24
environmental energy, 16
  see also conservation
epistemology: theory of, 74
  see also developmental epistemology; genetic epistemology
Eskimos: and language, 32
exercise: and energy, implied contradictions, 51-2, 67
experience: communication of, 32-3
experimental work, 93-4
exploration: about energy, 94-6
explosion view: of energy, 72-3

Faraday, Michael: and perpetual motion, 176
feeling: and action, 147-9
Feyerabend, Paul: 'anything goes', 23
Feynman, Richard, 119
fields of energy, 9
figures of speech: and energy, vii, 9
Flanders' Interactional Analysis, 64
food: chains, 51
  energy from, 55, 71-3, *passim* 75
force: as energy, 10, 12, 26-7, 43, 55, 67, 69, 101
  and *Kraft*, 34
  life, 44
  vital, 43-4
  see also living force; *vis viva*
Freud, Sigmund: and mental energy, 15-16
fuel: and conservation of energy, 125, 188
  fossil –, resources, 142-3, 144, 160, 188

Gaia: science of, 188
Galileo: and energy conservation, 123, 133
gender: responses about energy, 45-7, 148
generalization, 54-5
  on energy, 42
  success of pupils making, 56
general knowledge: adults', 3-4
  children's, 3-4
  citizens', 78-9
  and language, 33
  meanings, 16-17
  shared, 2
  nature of, 1-17, 98
  stock of, 65
  see also citizen knowledge; knowledge; meanings
generative learning model: *see under* learning
genetic epistemology, 19, 20, 73, 89
German language: and energy words, 34
girls: responses to energy, 45-7, 148
  see also gender
Guesne, Edith: and children's ideas on light, 22

Halliday, M.: and use of language, 33
harnessing: energy, 7, 177-8
  hydroelectric power, 15
Head, John, 147
health: and energy, 43, 47-8, 51, 53, 67
heat: children's ideas on, 22
  energy, 120, 134-5
  flowing, 178-80
Helmholtz, H.: and Conservation of Energy, 181-3
Hesse, Mary, 103
history: and colloquial speech of energy meanings, 7
  of energy technology, vii, 167-88
horsepower, 177-8
Humanities Curriculum Project, 154
human kinetic energy, 47-8, 51, 53, 67
hydroelectric power: harnessing, 15

ideas: and communication by language, 32-3
ideas, children's: changes, in, 29-30
  and culture, 36
  discussion of, 31, 36
  exploration of, 19, 22, 36
  and misconceptions, 26-31
  persistence, and commonality, 25
  and talking, 59
ideographic approach: to children's ideas, 22
implied contradictions, 51, 69
information: communication of, 184-5
inner speech, 65
intersubjectivity: of common-sense knowledge, 6
interview techniques, 19

Kelly, George: theory of personal constructs, 23

197

Index

Kelvin's mathematical scale of destruction, ix
kinetic energy, 83
  see also human kinetic energy
knowledge: and action, 89–91
  avoidance of, and risk perception, 87–8
  common sense of, 4–6
  disembedded, 80, 98
  embedded, 80
  everyday, 36
  formal, 144–6 *passim*
  informal, 144–6 *passim*
  meanings in, 76
  as power, 184–5
  scientific, 36
  social construction of, 75
  social stock of, 4, 73, 78, 103, 144, 160–1
  specialized, and technology, 186
  see also citizen knowledge; general knowledge; life-world knowledge; personal knowledge
Kraft: and energy, 34
Kuhn, Thomas: and paradigm shift, 28, 30
  and relativist philosophies of science, 21

language: and intellectual advance, 75
  as intrument of thinking, 62, 74
  learning, through use of, 75
  and perception, 32–6
  social functioning of, 74
laws: of effect, 19
  of energy, viii
  of recency, 19
  see also Thermodynamics: Laws of
learning: active, 37
  conscious, 92
  of correct science, 30
  general knowledge, 78–97
  informal, 79, 80, 96
  model, generative, 117–18
  need for, 78–9
    as motivation, 98
  rote, 37, 74, 129
  see also CLIS
Leeds University, CLIS project, 30–1
Leibnitz, von, German mathematician and philosopher: and *vis viva*, 10–11, 12, 123, 181
libido: and energy, 15

life: and energy, 9–11, 43–4, 183
  see also vis viva
life-world knowledge, vi, 50, 80, 83, 96–7, 136, 143
  and formal scientific knowledge, 98–115 *passim*
  crossing between, 112–15
light: children's ideas on, 22
  conception of, in drawing, 57, 58
  in Nigeria, 35
  misconceptions on, 26
living associations: and gender differences, 47
living energy: themes of, 47–8, 50
living force, 16
  see also vis viva
Locke, John, 53
logic: lack of consistency, in children's thinking, 21
  and rhetoric contrasted, 3

Malaŵi children's poems: on energy, 40, 41
mass: energy out of, 14
massless: – abstraction, 16
  energy, 9
matter, 14
Mead, G.H.: and communication of ideas, 32
meanings: blurring of, 16
  cluster, 42–4
  common, for energy, 7, 42
  comprehensible, 17
  in general knowledge, 2, 3, 76
  shared, 17, 62
  historical, 7
  mechanistic and political, 48–9
  and misconceptions, 28
  over-arching, 49–50
  in physics, 81
  – 'right', for tests, 108
  themes of, 47–8, 79
  variation, 53
  viable, 17
measurement: of energy, 4, 14, 54
memory, 169
  and tests, 108–10
mental: development, internal stages, 20
  energy, 14, 15
metaphorical redescription: scientific theory as, 103
metaphors: construction, by children, 38
  dead, 15

Index

and energy, 15–16, 103, 187
  hidden, 15
misconceptions: by children, 26–31
models: distanced, 38
  over-arching, 38
Monk, Martin: and children's constructs, 38
motion, 26–7
motor memories, 45
movement: and energy, 43–4, 50, 55, 101, 170
  as life force, 44
  running down, 122

Nature, scientific journal, 12–13
neo-Piagetian research, 20
Newton, Isaac, 9, 171, 179
  laws of motion, 12, 91
New Zealand school of research, 23
nomothetic attitude: and children's ideas, 22
non-renewable energy, 48
nuclear power: and risk perception, 87–8, 150–3, 156, 160

observation: theory-laden, 93
oil shortage, 143
open approach: in trails, 71
organization: and growth, 183–4
ozone layer: hole in, 152

paradigm shift, 28
passive energy, 94
people: energy of, 44–5
perception: and primitive notions, 31
perpetual motion, 173–7, 182
personal: knowledge, 6–7
  and risk, 157
  proto-scientific notions, 7
perspectives: exchange of, 162, 163
  modelling of, 162
petrol: energy in, 71
phatic communion, 63
phenomological primitives: *see* p-prims
photosynthesis, 28, 29
physical sciences: girls' study of, 47
physics: and change, 8
  and children's theories, 45
  and gender, 45–7, 148
  as general knowledge, 81
  teaching of, 98
Piaget, Jean: and children's ideas, 19–20, 38, 73–4, 89, 91, 104, 123, 124

Pines, Leon: and childen's misconceptions, 27
Planck, Max: and product of energy and time, ix
poems: on energy, by Malawi children, 40, 41
potential energy, 73, 74, 82, 83
power: as energy, 10, 11–14, 43, 55, 66, 68, 172, 173
p-prims (phenomenological primitives), 31, 45, 54, 91
primitive notions, 31
Principle of Conservation of Energy: *see under* Conservation of Energy Principle
probability: understanding of, 152
problem-solving talking during, 62–3
prose: about energy, by children, 41–2
proverb, common, 183
  and alogical rhetoric, 53
pseudo-discussion, 62
public energy debate, 49
Public Understanding of Science, vi, 85, 87, 141, 151, 152
pupil paradigms, 22
pure energy, 8

quality, 53
  measurability of, 54

rationality: and change, 30
realism, 104
reaping benefit, 7
recall, 149
  and tests, 108–10
relativist philosophies: in science, 21, 24
renewable energy, 48, 142, 143
research: action, in classroom, vi, 130, 136–40
  programmes of, 30
  answers, 36–7
  on learning, 26–8
  neo-Piagetian, 20
  new educational, 20–1
respiration, 28
reversibility: of energy, 123–5, 136
rhetoric, 2
  alogical, of everyday life, 53
  and logic contrasted, 3
'rights': of people, 163
risk perception: balancing, 164
  and energy knowledge, 87–8, 151–3
  quantifying, 159

199

Index

and statistics, 152
  students' assessment of, 157–61
rote learning, 37, 74, 129
Rumford, Count, 180
running down principle, 131–3, 134, 146

Sapir, Edward: Whorf hypothesis, 33–4
science: amateurs in, 83
  education, 142
  indicators, 142
  influence, on life, 83
  knowledge, individual and personally constructed, 25–6
  new approach to, 21
  popularization, on television, 14
  public understanding of, vi, 85
  relativist philosophies, 21, 24
  and social values, 141–66
  teachers of, x
  and children's misconceptions, 26
Science Policy Support Group, 156
scientific: concept, 53–4
  growth and change in, 36
  ideas, and use of verbs, 55, 56
  journalism, 12
  knowledge, 79, 160–1
    formal, 98–115
    and life-world, 99–115
    crossing between, 112–15
  measurement, trust in, 151
  processes, 85–7, 151
  societies, 12, 83, 182
scientists: bias of, 162
  consensus among, 149–50
  power of invention, 143
shortage commodity: of energy, 49
similes: children's use of, 38, 105–6
  close, 38
  descriptive, 38
  and energy, 14–15
Simpson, Mary: and children's misconceptions, 28
Skilbeck, Malcolm, 141
social issues: and energy knowledge, 141–66
solar energy, 145
source: of children's knowledge, 96–7
  of energy, 70–3
speech: colloquial, and meaning of energy, 7
  figures of, and energy, vii
statistical mechanics, viii
statistics: use of, 151–2
steam power, 173, 178

Stevinus, Simon: and perpetual motion 174, *175*
Stewart, Balfour: and energy of character, 13
storage: of energy, 70–3, 93, 94, 121, 136
structuralism, 31
sun: energy from, 50–1, 57, 69
  see also solar energy
Surrey, UK, School of Research, 23
syllabus: Delphi method of design, 149–50
synonyms: for energy, 42
systems: in physics, 100, 102, 105

tacit knowledge, 91–2, 96
talking: during problem-solving, 62–3
  learning through, 64
  as research tool, 59–77
  unconstrained, 65
tape recorders: use of, 64
teachers: of science, x, 25
teaching: and action research, 116, 118, 130, 137–40
  about conservation of energy, 116, 126–40
  model of, 116
  changes in, 129–30
  and two domains of life-world and scientific knowledge, 99–115
  models of, 141
  of energy, 93
  methods of, 120
technical knowledge, 78, 79, 88–9, *90*
technology: adaptability, 168
  defining, 168–9
  in education, 167–8
  history of, 170–85
  internalization, within culture, 186
  new, 169
  philosophy of, 185–8
teletransportation, 9
television: and adult's concerns, 49
  and popularization of science, 14
  and public understanding of energy, vi, 143, 154, 156
  space sagas, and massless energy, 9
temperature: children's ideas on, 22, 92
  and elderly, 87, 88
tests: lack of success, and children's understanding, 107–8
theoretical fluids, 179
Thermodynamics: Laws of, 125–6, 187

Index

reformulating Second Law, 133–6
running down, 146
thinking: children's, 21
  untutored, 79
  different ways of, 18–20
  disembedded, 79
  embedded, 79
Third World children: and language of energy, 34–5
time: and energy, ix
Tiv language, Nigeria: and energy, 35
Todd, Francis, 63
trails, 69–73

validity analysis, 138–40
value-free activity, 18, 19
values, 164
  and feeling, 147
  and social justice, 165, 166
verbalizers, 57
verbs: to define scientific ideas, 55, 56
Viennot, Laurence: work on force and motion, 26–7, 113
vis mortuus: dead force, 11
visualizers, 57
vis viva: living force, 10–11, 12, 180
vital force: and energy, 43–4

vitalism: theory of, 11, 180–3
volume, 54
von Helmholtz, Hermann, 83
Vygotsky, L.S., 62, 75, 92, 132

Wason, Peter, 130
wasting energy, 84, 133, 134, 148, 173
water wheels, 171–2, 174, *175*
weight, 54
Wellington, J.J., 155
wheels, 171–2
  see also water wheels
Whorf, Benjamin: and Sapir hypothesis, 33–4
windmills, 171–2, 174, *175*
Woolnough, Brian, 93
word clusters, 42–4
words, for energy: associations, 36, 44
  concept, 34
  in German, 34
  in physics, 81
  popularity of, 16
  as symbol, 16
work: abstract concept, 100–2
  as energy, 11–14, 55, 69, 71, 82
  units, 14, 102
'write about energy', 41–2